So You Have to Teach Math?

Sound Advice for Grades 6–8 Teachers

CHERYL RECTANUS

MATH SOLUTIONS PUBLICATIONS
SAUSALITO, CA

Math Solutions Publications
A division of
Marilyn Burns Education Associates
150 Gate 5 Road, Suite 101
Sausalito, CA 94965
www.mathsolutions.com

Library of Congress Cataloging-in-Publication Data

Rectanus, Cheryl.
   So you have to teach math? : sound advice for grades 6–8 teachers / Cheryl Rectanus.
      p. cm.
   Includes bibliographical references.
   ISBN-13: 978-0-941355-73-5
   ISBN-10: 0-941355-73-X
  1. Mathematics—Study and teaching (Middle school) 2. Mathematics teachers—Training of. I. Title.
   QA11.2.R434 2006
   510.71′2—dc22   2006017044

Editor: Toby Gordon
Production: Melissa L. Inglis
Cover design: Leslie Bauman
Interior design: Angela Foote Book Design
Composition: Interactive Composition Corporation

Printed in the United States of America on acid-free paper
10   09   08   07   ML   2   3   4   5

A Message from Marilyn Burns

We at Math Solutions Professional Development believe that teaching math well calls for increasing our understanding of the math we teach, seeking deeper insights into how children learn mathematics, and refining our lessons to best promote students' learning.

Math Solutions Publications shares classroom-tested lessons and teaching expertise from our faculty of Math Solutions Inservice instructors as well as from other respected math educators. Our publications are part of the nationwide effort we've made since 1984 that now includes

- more than five hundred face-to-face inservice programs each year for teachers and administrators in districts across the country;
- annually publishing professional development books, now totaling more than sixty titles and spanning the teaching of all math topics in kindergarten through grade 8;
- four series of videotapes for teachers, plus a videotape for parents, that show math lessons taught in actual classrooms;
- on-site visits to schools to help refine teaching strategies and assess student learning; and
- free online support, including grade-level lessons, book reviews, inservice information, and district feedback, all in our quarterly *Math Solutions Online Newsletter*.

For information about all of the products and services we have available, please visit our website at *www.mathsolutions.com*. You can also contact us to discuss math professional development needs by calling (800) 868-9092 or by sending an email to *info@mathsolutions.com*.

We're always eager for your feedback and interested in learning about your particular needs. We look forward to hearing from you.

Math Solutions®
PUBLICATIONS

Contents

Acknowledgments

Writing for me is not a solitary pursuit. Many people have contributed to this book by sharing their experiences and expertise, critiquing drafts, nudging the manuscript along, and providing inspiration, encouragement, and support.

I am grateful for the many students and teachers with whom I've worked for the past twenty years. Thank you also to Pam Wilson, Genni Steele, and Judy Koenig, and to my colleagues with Portland Public Schools, Marilyn Burns Education Associates, and Teachers Development Group.

Special thanks to Toby Gordon and Joan Carlson, for reading every word of the manuscript and providing critical feedback.

I'm especially indebted to Cathy Boyce, Margaret Calvert, Susan Stein, and Ruth Tsu. Learning together as we inquire into our practice has deeply impacted my work, and thus this book.

Thank you to my parents for their love and support, and to my grandmother, a teacher who nurtured me during my first years in the classroom.

Finally, I reserve special gratitude for Fred Rectanus, for more than I can say.

Chapters

and

Questions

Seven   *Using Manipulative Materials*

Eight   *Using Calculators and Computers*

Introduction

Teaching middle school mathematics is a complex endeavor. Adolescents experience significant intellectual, physical, emotional, and social changes. In addition, today's diverse student bodies necessitate that you connect and build relationships with students whose ethnicity, beliefs, and socioeconomic status may differ from your own. The middle school math teacher is thus presented with particular challenges. This book is meant to address those challenges and help give you support as you work to create a safe, encouraging, and equitable classroom; as you aim to be responsive to the needs of both advanced and struggling students; and as you strive to be a thoughtful and effective math teacher.

Whether you're right out of college, have been teaching for a few years, or have been in the classroom for most of your career, you probably have questions about the math that is appropriate for middle schoolers as well as how to effectively teach it. Questions like the following:

- *What do I need to include when considering my instruction for the year?*
- *How do I set up a learning environment in which students are willing to take risks?*
- *Sometimes when students work together, one or two individuals do most of the work while the other members goof off. What are ways to ensure individual accountability when my students are working with a partner or group?*
- *I have a number of students who are just learning English, and they struggle with the writing I assign in math class. How can I best support them?*
- *What are some effective ways to connect with parents of adolescents?*
- *I know that my less-experienced learners will benefit from using manipulative materials, but do my most-experienced learners really need them?*
- *What are the benefits of using calculators and computers in math class?*

This book addresses these questions and many more. No matter your interests, prior experiences, or particular dilemmas, there is plenty of help here for anyone who wants to be successful teaching mathematics to adolescents.

Adolescents' Intellectual Needs

As children grow into adolescence, they think more abstractly. The process doesn't take place at the same time for each child, especially when you consider that students' understanding is influenced by their opportunities. For the teacher trying to provide responsive math instruction, this uneven transition raises challenging issues. Topics that involve proportional reasoning, for example—including the concepts of ratio, slope, similarity, scaling, linear equations, percents, probability, interest, and rates—can be difficult even for the most adept students. Classrooms include students with a wide range of abilities and experiences, and middle school math teachers are faced with the issue of effectively meeting many different needs. In addressing skill disparities among their students, middle school math instructors regularly deal with the following concerns:

- *Some of my students really struggle with math. What are some ways to make sure that I meet their needs?*
- *How do I address the needs of students who finish assignments quickly and who are ready to tackle more complex work?*

Adolescents' Social and Emotional Needs

Adolescents have pronounced social and emotional needs. Middle school students have obvious differences in their levels of maturity, and their behavior reflects these differences, which has earned them a reputation for being difficult to teach. "Scattered," "disrespectful," and "self-absorbed" are several of the ways I've heard sixth-, seventh-, and eighth-grade students described—I was actually asked once how I got "stuck" teaching middle schoolers. In my opinion, however, adolescents' negative reputation is undeserved. My experience tells me that students in middle school are interesting, compassionate, respectful, fun loving, and eager to learn. Are they sometimes trying and exasperating? Of course. Teachers who bring patience, understanding, and skill to their practice will find teaching mathematics to adolescents a joyful and rewarding experience. To get the best from students, teachers will want to consider questions like these:

- *How can I support students in working together productively?*
- *Students who make mistakes during class discussions are sometimes laughed at or taunted. How can I prevent this from happening?*

- *Students' contributions are sometimes dismissed or marginalized by others in their small group who are perceived as being "smarter" or more popular. How can I address this?*
- *How do I establish a classroom atmosphere in which class discussion is valued? How do I help students learn respectful norms for interaction during discussions?*

My Journey

While I now find it immensely satisfying to teach mathematics to middle school students, this wasn't always the case. I started my career teaching young children and gradually moved up the grades. I loved teaching elementary students, but I found it difficult to do justice to all the curricular areas for which I was responsible. I felt like a jack-of-all-trades and a master of none. It was the desire to specialize in mathematics education, more than a compelling interest in adolescents, that propelled me to enter the world of middle school math teaching. I found it a fascinating place.

That first year in middle school was an eye-opener. I was impressed with the sophisticated thinking my students exhibited as compared with my elementary students, as much as I appreciated the fact that they had not yet quite outgrown childhood. One moment they'd be writing algebraic rules for working on a coordinate grid and the next they'd be asking me if they could play *Musical Chairs*. I found myself going to my colleagues' classrooms during lunch and after school to mull over all sorts of issues. We discussed the roles of discussion and writing in math class; we debated about how to tell from their work what students knew and understood and what the implications were for our instruction; we shared tools for assuring that all students—regardless of their socioeconomic status, cultural background, language, personality, and achievement levels—enjoyed equitable participation in small groups and were accountable for success in learning mathematics; we talked about efficient ways to read and score student papers. Many of the following questions emerged during the course of our regular discussions:

- *How often should students use manipulative materials? Should they be used daily?*
- *How can I help my students become more proficient with their basic facts so they don't always grab a calculator to compute problems they should be able to figure mentally?*
- *Should I grade or score every assignment?*
- *How should I prepare students for the tests that the district or state requires?*
- *What if parents tell me that they don't understand the math behind their child's classwork or homework?*
- *Getting students to complete their math homework is difficult. What are ways to encourage students to complete assignments?*
- *What ingredients are necessary to support adolescents' writing in math class?*

In taking on these and other questions in this book, my intention is to stir your thinking about the decisions you make as a teacher and, as a result, to promote the mathematical learning of all of your students. I hope to give you support, encouragement, and direction—as well as lots of food for thought.

How to Use This Book

I recommend that you reflect on what I've written to consider how it might impact your mathematics teaching. Whether you disagree with an idea or suggestion or find yourself in favor of a particular pedagogical position, examining and questioning why you have a specific response will help to make some of your assumptions, values, and beliefs visible, allowing you to reflect on them, weigh various points of view, and enhance your professional growth.

I also suggest that you discuss the ideas in this book with your colleagues. You might consider beginning a faculty book-study group in order to carve out time and space to collaboratively reflect on your teaching practice. Participating in a professional learning community with one's peers is associated with increased student learning and achievement. It's worth your time and effort to meet regularly with other math teachers in your school and district.

A Final Word

All decisions about teaching have benefits and limitations, as do the answers in this book. My recommendations and suggestions are based on many years of experience teaching middle school mathematics, facilitating professional development for teachers, working with individual teachers and math teams in schools, and thinking deeply with my colleagues about how to provide an equitable education for all students. But you'll need to take your specific teaching reality into consideration as you decide what to embrace and what to fine-tune so that it fits for you. My hope is that you are able to use what I've learned to support your continually developing expertise and effectiveness as a middle school mathematics teacher.

One

Preparing for a Successful Year

Cameron and Shelly both teach eighth-grade math and want to be sure that their middle school students are well prepared for high school math. Their district has just published new math standards, and the district testing requirements are new as well. Cameron and Shelly aren't sure about how these changes should impact their math teaching. What should they be thinking about in terms of instruction? What preparation will students need for the new assessments?

Max has different questions about the upcoming year. He has been teaching in a self-contained sixth-grade classroom for five years and has just been reassigned to eighth grade. He enjoys teaching math and knows the sixth-grade math curriculum well, but he hasn't taught eighth grade since his student-teaching assignment. He feels he needs to determine the year's goals and learn what topics his eighth graders should study in mathematics and when they should be taught.

Madison has other concerns. She's getting ready to teach sixth grade and wants to know what models the fifth-grade teachers used in their instructional program. Madison knows that reminding students about models or strategies that they learned in an earlier grade helps them connect their current learning with their prior knowledge. She also wants to meet with the math teachers in her school to discuss how topics are being built on from one year to the next.

As these teachers' concerns illustrate, there are many things to consider when preparing for a year of math teaching, from standards and assessment to curriculum and instruction. Looking at the overall picture can help you identify issues that are important for you to address in your instructional program.

1 What's the best way to determine what math concepts and skills I need to teach for my grade level?

It's wise to have a governing idea about the year's math instruction. You need to comprehend the big picture of what students in your classes should understand and be able to do mathematically by the end of the year. This can be a bit tricky if you are a new teacher or a teacher new to a particular grade level. You may remember what it was like learning math as a student in the grade level(s) you're teaching, but what and how it was taught is likely to have changed. Take advantage of several types of resources rather than relying solely on one source for determining your teaching goals. Each will inform your thinking by offering you a different perspective.

You might start with the national standards, which provide overall guidance about the elements of a strong math program, including content standards and process standards—the "what" to teach and the "how" to teach it. Don't expect to find daily lesson plans in the national standards document, but it will provide a helpful picture in which to frame your more specific grade-level concerns. *(See Question 2 for an overview of the national standards.)*

Next, check your state standards. Many states have patterned them after the national standards. In addition, many states include the content of their standards on statewide assessments, so you'd be wise to familiarize yourself with them.

To provide more specifics, many schools and school districts have developed a set of local math standards. These are important to consult when developing an overall picture for the year. Variously called math frameworks, curriculum guides, teaching and learning standards, district benchmarks, student learning objectives, and performance standards, they provide information about what students should learn. Read these carefully. Look at the math concepts and skills in earlier and future grades to understand the context for the math you teach. It's important to know what students learned in previous years and what will be covered in the next year or years. If you look only at the standards for your grade, you may miss the big picture in planning your instruction. For example, a new sixth-grade teacher read just his district's sixth-grade math standards and was concerned when he found that integers were not addressed. Responding to this perceived lapse, he developed and taught a unit on integers for his students only to learn that integers were thoroughly covered and assessed in seventh grade. Furthermore, in teaching his unit on integers, he ran out of time to teach about probability, which *was* included on the sixth-grade state assessment.

Your school or district may not have a specific set of math standards, using instead the documents produced by your state. Ask one of your math colleagues or your principal for information about obtaining copies of various standards documents, or check your district or state education department's Web site.

Finally, the instructional materials provided by your school or district for teaching math will include specific information about planning the year's instruction. Lessons may be described in detail, helping you think about and prepare the materials needed for each learning segment. As you become more experienced in teaching math, you'll likely supplement your curriculum materials with activities that you've learned about from a variety of sources: colleagues, workshops, teacher resources, and your own experiences.

2 What should I know about the national math standards?

Principles and Standards for School Mathematics, our current national math standards, is published by the National Council of Teachers of Mathematics (2000c). This document sets forth a vision for school mathematics that includes the following ideas:

- All students (including those traditionally underrepresented in mathematics) can learn math when they have access to high-quality math instruction.
- Our world is constantly changing and thus the need to understand and use math in everyday life is increasingly important. This includes the need to understand quantitative information and the ability to problem solve in and out of the workplace.
- Learning math today requires much more than memorization of facts and procedures. Students must be able to think, reason, and solve problems with depth and understanding.
- Students learn math when they are actively involved in sense making.

Six guiding principles—equity, curriculum, teaching, learning, assessment, and technology—are described in the first part of the document. These are followed by standards for four separate grade-level bands: pre-K–2, 3–5, 6–8, and 9–12. Each grade-level band provides specific information about ten standards (five content standards and five process standards) as well as suggestions for classroom instruction.

The content standards include:

Standard 1. Number and Operations
Standard 2. Algebra
Standard 3. Geometry

Standard 4. Measurement

Standard 5. Data Analysis and Probability

The mathematics that students must learn is defined in the five content standards, which describe "what" we teach. Note that there are overlaps among the content areas. For example, as you might guess, number appears in all of the content standards. Algebra and geometry share content with respect to patterns, functions, and spatial sense. Regardless of the instructional materials you use, the content standards should all be addressed. They give you a lens through which to think critically about your curriculum materials.

The five process standards describe the "how" of bringing the content standards to life for your students. They include:

Standard 6. Problem Solving

Standard 7. Reasoning and Proof

Standard 8. Communication

Standard 9. Connections

Standard 10. Representation

All ten standards are necessary in a comprehensive and coherent math program.

3 My instructional materials give me the direction I need for planning day-to-day lessons, so why is it important to consider national or state standards?

While your instructional materials give day-to-day support for teaching your students, national and state standards provide an overall picture of math teaching and learning that can inform your thinking as you make your daily instructional plans.

An analogy can help illustrate this idea. Say you're planning a garden. You might decide on specific plants to put in it and determine where in the garden each will go. But what if you get to the nursery and the specific plants you want are unavailable? You'd be better prepared if you had a broader sense of what you're trying to achieve in the garden, the kind of soil you have, the amount of sunlight it receives throughout the day, and so forth. Knowing these general characteristics of your garden plot will enable you to achieve success with the individual plants contained in it. In the same way, national and state standards can be thought of as the broad general characteristics—the garden plot—that provide a context for specific instructional materials.

When you're preparing to teach any lesson, there are many things you need to take into consideration in order for the lesson to be a success. You need to think about the objective for the lesson, the context and materials needed, how students will be organized—in small groups? with a partner? alone?—how you'll introduce the lesson, explain your expectations, support and challenge students as they work, figure out how to deal with students who finish early, and so on. Sometimes it's so easy to get caught up in planning for these important details, you forget the larger purpose and mathematical goals of the lesson and where these ideas fit into students' overall math learning. Standards can help with that, and they are especially useful in helping you think of extensions for your lessons. If you know where the lesson is going mathematically, it's a lot easier to think about what questions and extensions are appropriate for your students.

4 How might I think about scheduling in all of the topics I need to teach, and matching my curriculum, major projects, and assessments to scheduling constraints?

It's helpful to map out your instructional year using your school or district calendar. This way you can take into consideration when quarters, trimesters, or semesters end, and when standardized assessments, holidays, and vacations are scheduled, all of which will help you make decisions about the order in which to include units and assessments.

The math team at one school I know gets together in late spring to plan out the next school year. Their school is on a quarter system, so they've learned that they can teach about two four-week units or chapters per quarter for a total of eight per school year. The teachers look carefully at when their school administers standardized tests, which is usually about two-thirds of the way through the school year. They conceive of the instructional year as beginning the day after standardized testing ends. They've learned that they cannot address every topic on their standardized test before it is administered in the spring, so they consider the topics taught after testing as preparation for next year's test.

The teachers place a greater emphasis on number concepts when their students are in sixth grade, but by the time students reach eighth grade, they are studying fewer number topics and more algebra topics, consistent with national recommendations.

Finally, when they plan the year, they think about unifying ideas that tie together the sixth- through eighth-grade topics, especially proportionality, which integrates many areas of middle school math, including ratio and proportion, percent, scaling, similarity, slope, linear equations, relative-frequency histograms, and probability.

A three-year plan that includes when major projects are assigned might look something like this:

	Fifth Grade	Sixth Grade	Seventh Grade	Eighth Grade
Quarter One		Number Theory: Primes, Multiples, Prime Factorization, etc.	Algebraic Thinking: Multiple Representations, Variables	Linear Relationships
		Data & Statistics: Measures of Central Tendency (Project)	Similarity (Project)	Geometry: Pythagorean Theorem (Project)
Quarter Two		2D Geometry: Polygons, Angles, etc. (Project)	Proportionality & Scaling	Algebraic Thinking: Exponential Functions
		Rational Numbers: Concepts & Relationships	Integers	Algebraic Thinking: Quadratics
Quarter Three		Measurement: Area & Perimeter	Data & Statistics: Large Numbers & Sci. Notation (Project)	Algebraic Thinking: Language of Algebra, Properties
		Introduction to Algebraic Thinking	Geometry: Surface Area & Volume	Geometry: Transformations (Project)
		Standardized Testing	Standardized Testing	Standardized Testing
Quarter Four	Introduction to Rational Numbers	Rational Numbers: Computation & Estimation	Rational Numbers: Review & Application	Data & Statistics: Box & Whiskers, etc.
	Probability: Experimental	Probability: Experimental & Theoretical	Probability: Area Models, Expected Value	Probability: Combinations

The teachers include the fifth-grade teachers when they plan, to ensure continuity. Having all the grades represented not only helps teachers to coordinate the topics taught and the order in which they are presented, but also enables them to articulate the models and strategies used in each grade level so that they are able to help students make connections from one grade and math topic to the next.

Two | Planning for Instruction

Eleanor, Yoshiko, and Michael were talking in the hall the week before school began. "Even though it's only my second year of teaching, I feel like I have a pretty good handle on the mathematics my students need to learn," Eleanor told her colleagues. "Still, I wonder about more effective ways to support my students who struggle with reading the math text, as well as those who need additional help in developing their number sense."

"Those are important things to consider," Yoshiko concurred. "I'm rethinking how I manage the classroom so that students will get the most out of each lesson."

Michael listened thoughtfully and suggested, "Let's set aside some time to talk about how we can help each other with our different questions. I'd like to hear your ideas about streamlining the curriculum. The text has so many problems in it, and some don't seem worthwhile. If we plan together, we can figure out which problems really engage students in doing mathematics and which might be omitted." The three teachers agreed to meet the next afternoon to support one another as they planned for a new school year.

Planning for instruction is vital if teachers want their students to become competent and confident math learners. Planning encompasses ideas such as determining a logical structure for your daily math class, acquainting yourself with the prior year's curriculum so that what students know informs your lesson plans, and becoming a critical consumer of your instructional materials. The more you are able to plan for instruction with your colleagues, the better. In thinking through mathematics teaching, this is especially true. Just as students benefit from being able to share their thinking with one another and hear new or different points of view, teachers gain from quality interactions with one another. Your math colleagues can share their experiences with respect to teaching the topic, alert you to things you might not have considered, and suggest ways to strengthen lessons. Furthermore, many studies show that participating in a professional collaborative learning community of one's peers is associated with increases in student learning and achievement (Newmann and Whelage 1995; Louis, Kruse, and Marks 1996; Talbert and McLaughlin 1993).

7

5 What do I need to include when considering my instruction for the year?

There are three major things to think about when getting ready for the school year: planning, preparation, and management.

First let's consider planning. Think of this as your road map for teaching. You begin with the desired destination—what you want your students to learn. Then you decide how to get there—what route to take, and the vehicle that will best support students on their journey. This involves thinking about what lesson will help students reach the mathematical goal, how you'll introduce the lesson and connect it to what students already know, how you'll assess student learning, and so forth. Many teachers find it helpful to write down specific directions or notes so that they have something to refer to during the actual lesson. It helps to reread your notes or plans prior to teaching the lesson, to make sure that they make sense to you. Is there anything missing? The more your plan helps you anticipate what might happen as you teach the lesson—both intended and unintended outcomes—the better.

Next consider preparation. While the plan is the "what" of a lesson, preparation can be thought of as the "why." Preparation requires careful consideration of the overarching mathematical goals for students as well as the specific goal for the lesson. When you prepare for a lesson, ask yourself the following kinds of questions:

- What mathematics is being taught? What do you expect students to get out of the lesson? You want to be clear about the point of the lesson and the essential mathematics in it so that you can help students focus on the mathematical target as the lesson unfolds. Some teachers craft a student-friendly question that is recorded on the board or overhead projector at the beginning of math class. The question is designed to communicate to students and visitors alike what the students are expected to learn. Examples include, *"What is a factor?" "How are patterns represented in coordinate graphs and how can you use a pattern to make predictions?" "What are ways to describe what's 'typical' in a set of data?"* At the end of the lesson, students can be asked to discuss and then write about their insights in responding to the question.
- How will the lesson connect to what students already know, both in terms of their general prior learning and specific models that they may have used in previous years? It's helpful to know generally what topics the students have been taught and what they were expected to learn. You can consult with your students' former math teachers or, if this is not possible, ask your students, either individually or during a class discussion, about their previous experiences and models they may have used.
- Given the questions you pose to them, how might students respond? What might their answers tell you about their mathematical understanding? What will you do if they appear to struggle? What happens if they easily grasp the concept? How will you

challenge their thinking in order to keep the mathematical trajectory of the lesson moving forward? As in planning, it helps to anticipate student responses. At the same time, you want to stay open to students' thinking; don't be so caught up in your expectations of their responses that you miss what they actually do say! Be sure to anticipate surprises. Even after many years in the classroom, I regularly encounter students who come up with unique ideas and strategies that I didn't consider despite my careful preparation, which is one reason I find teaching math to adolescents so interesting.

Management is the third aspect to consider when getting ready for the year. This is where the details of organization come into play. For example, if you're teaching a lesson that requires the use of graphing calculators, you don't want to go hunting for them in the closet at the beginning of class only to find that the batteries are dead. Get the calculators out and make sure they're working properly before class starts. There is no end to the mischief that adolescents can get into when their teacher is distracted (not to mention the wasted instructional time). The same goes for other materials that students might need for a lesson. Consider how best to distribute them to minimize disruptions. Likewise, think about how to organize students for the lesson. Will they work alone? With a partner or small group? How will this occur? *(See pages 36–43 for information on grouping students.)* What about accountability for their work and behavior? What will students turn in at the end of the lesson? What about homework? *(See Chapter 11 for in-depth information on homework.)*

While elements of planning, preparation, and management overlap, it is essential to think through all three when getting ready for the new school year. I find it invaluable to consult with a colleague or my math team when thinking about the year's instruction. My friend Ruth reminds me that "no one of us alone is as smart as all of us together." Getting feedback from colleagues is especially useful when preparing to teach something that is hard to teach, difficult to learn, or an important mathematical idea. Ideally, teachers will have an opportunity to observe the lesson as it plays out in another teacher's classroom; they are then able to collect data about students' reactions, responses, and learning that can inform their own teaching. If this isn't possible, ask a trusted colleague to come into your classroom to observe a lesson that you've planned and give you feedback about its effectiveness.

6 My textbook appears to be missing some concepts that the district considers important at my grade level or that appear on the standardized tests. What's the best way to handle this?

First, go over the information in your state and local math standards so that you are sure about what students are supposed to learn in the grade level that you teach. You want to be certain that the material not included in the textbook should actually be taught at the grade level in question. Second, take the time to learn about the curricula you, your school, and your district are using. You have to understand how your curriculum is structured in order to teach it effectively. The concepts in question might be embedded in the materials rather than taught explicitly, or covered in a later unit. Keep in mind that most curriculum materials are published for a broad audience and, therefore, state and local standards may not align perfectly with the instructional materials. If you don't see a particular concept or skill in your grade-level materials, ask experienced teachers and curriculum-support specialists in your school or district where it is addressed. Many districts now align their instructional materials with math standards for each grade level so that teachers can see what will need supplementing.

If you determine that something is missing from your instructional materials, consult with your colleagues to learn how they are addressing the issue. Many times they'll have good ideas for lessons and can share their experience in teaching them. If not, consult the wealth of teaching materials available for grades six, seven, and eight.

7 My textbook covers too much material. If I taught every lesson included in the book, I would need two years rather than one! What are some ways to combine chapters or topics to make the amount of material more manageable?

It's not uncommon for textbooks to be crammed with topics. Publishers are trying to meet the needs of a variety of customers, and it's safe to say that most textbooks err on the side of including too much material rather than not enough. This is especially true when you consider the plethora of supplemental materials that many publishers include in their text series. A major problem with including too much material is that the curriculum is at risk of being "a mile wide and an inch deep." In other words, a large number of topics are covered superficially and quickly.

It helps to look at your materials with a critical eye after becoming familiar with your local standards. What sections are superfluous? redundant? not taught at your grade level? Leave those out. Are there pages that include forty problems when ten well-chosen problems would suffice? Assign only those ten. As you work your way through the text, consider using sticky notes in two different colors to distinguish between those problems and activities you want to include and those you want to omit. Using sticky notes also allows you to jot down ideas and comments for future reference. *(See Question 10 for information about maximizing the instructional potential of various problems or tasks.)*

8 My instructional materials for students require them to do a lot of reading. What are some tips for supporting students who have difficulty with reading?

Some instructional materials today are "text dense," or packed with words that students must comprehend in order to make sense of the mathematics they are learning. Students typically don't have difficulty decoding the words, but many struggle with comprehension. There are several techniques you might try:

- Read the problem aloud and think about it as a class. You may read the problem yourself or have a skilled student do so. Next, discuss what the problem is asking students to do. It helps to have several students offer explanations, so that the same idea is expressed in different ways. Be sure to highlight unfamiliar vocabulary and discuss difficult words. While going over the problem as a class is a useful technique, teachers should be careful not to give away methods for solving the problem when helping students understand what's being asked.

- Have students work together in partners or small groups. Sometimes adolescents who struggle with reading comprehension are more willing to talk with a classmate about the problem in a relatively private setting than to ask for help in front of the whole class. As you circulate through the class, you can check in with students who you know may run into difficulty with the reading.

- Have students make T-charts, in which they jot down what they know on one side and what they need to find out on the other. I began using this technique as a way to help uncover what it was that confused my students. Students might admit to not understanding something, but it can be difficult to determine where the difficulty lies—if they know the math but are confused by the reading or if they understand the question but don't know how to approach the problem mathematically.

- Encourage your struggling readers to slow down. Many times adolescents worry about appearing stupid in front of their peers and the teacher and so they rush—guessing at what is called for and applying strategies or operations without thinking about the numbers they're using or even if the approach makes sense.

- Write an abbreviated version of the problem on the board or overhead. While this technique can be useful, you have to be careful that you don't diminish the intended cognitive level of the task.

• For nonnative English speakers, consider having the materials translated. This requires advance preparation but can provide huge benefits for students who are learning English. *(Question 70 in Chapter 6 provides more information about supporting English-language learners.)*

9 My instructional materials use a problem-centered approach to teaching math, and skills are embedded in the curriculum. Don't students need to know the basics in order to be successful with such materials?

First, let's talk about what is meant by "the basics" with respect to arithmetic. Some understand this as referring to adding, subtracting, multiplying, and dividing whole numbers. Others expand that definition to include addition and multiplication facts committed to memory (addition facts are sums with addends from 1 to 10; multiplication facts are products with factors from 1 to 12). Let's think about middle school, however. In this case, the basics can be thought of as the ability to efficiently and accurately add, subtract, multiply, and divide with whole numbers, fractions, decimals, and percents using any algorithm or procedure that makes sense to the student. I would argue that students who know the basics, in addition to having computational proficiency, know how to apply computation skills, including:

1. choosing the operation needed;
2. choosing the numbers to use;
3. performing the calculation using mental math, paper and pencil, or a calculator;
4. evaluating the reasonableness of the answer (which requires being able to estimate).

So applying computational skills to solve problems and developing good number sense are essentials for math learning. This leads me to the next part of the question—the assumption that students who struggle with basic skills cannot be successful using a problem-solving curriculum. Some teachers believe that if students cannot add, subtract, multiply, and divide with fluency, they cannot successfully engage in such a curriculum. I don't think that this is true.

Consider learning a sport such as skiing. Say that you knew nothing about the sport but were taught all the basic skills you needed to know before you got on the slope. You learned about the laws of gravity, how to put on your skis and orient them to slide and stop when going downhill, how to plant your poles during turns, the correct posture to employ, when and how to bend your knees, and so forth. It makes no sense, obviously, to learn the skills without the context—without actually being on the slope. Similarly, in math class it is important to help students develop an understanding of concepts and solidify and use basic and other skills in the context of problem solving. Someone once shared with me the following:

Skills without understanding are meaningless.
Understanding without skills is inefficient.
Without problem solving, skills and understanding have little utility.

I think this is a useful way of thinking about the importance of all three areas of math—understanding concepts, skills, and problem solving. All are necessary and one does not necessarily come before the others.

10   How can I open up my skills-centered instructional materials and use them to support students' mathematical understanding?

Stein and colleagues (1999) analyzed hundreds of lessons during a five-year period and noted that gains in student learning were greatest in classrooms in which instructional tasks consistently encouraged high-level student thinking and reasoning and least in classrooms in which tasks were consistently procedural in nature. Thus it's important to analyze and think about what each problem or lesson requires of students and, in some cases, to enrich the task so that it better supports the development of students' mathematical understanding. For example, consider the following problem: *Imagine a row of one hundred equilateral triangles in which the length of each edge measures one unit. What will the perimeter measure?* The focus in this problem is on producing a correct answer. But students have to do a lot more thinking if I ask them to demonstrate their mathematical understanding of the problem using this extension: *Imagine a row of one hundred equilateral triangles in which the length of each edge measures one unit. What does the perimeter of the row measure? What would the perimeter be for any number of triangles? Represent your thinking in two or three ways and explain how the representations are connected.* This question engages students on a different level, challenging them to make meaningful connections between the problem and the mathematical reasoning required to solve the problem. It's useful to consider the kind of mileage you can get out of any problem or task with respect to the development of students' mathematical thinking and reasoning. Sometimes you'll elect to implement a task as written, sometimes you'll modify a problem to make it more cognitively challenging for students, and other times you'll decide not to use a task or question because there's just not enough mathematical "meat" in it.

11 I hear a lot about the importance of number sense. What exactly is number sense? And how does it relate to the basics?

Number sense can be thought of as "refer[ing] to an intuitive feeling for numbers and their various uses and interpretations; an appreciation for various levels of accuracy when figuring; the ability to detect arithmetical errors; and a common sense approach to using numbers" (Reys 1991, 3). Students with good number sense understand the following concepts:

- *Relationships between and among numbers.* Students can break numbers apart and put them together in various ways. For example, to calculate 96×70, a student with solid number sense might first multiply 90 by 70 to get 6,300, next multiply 6 by 70 to get 420, and finally add together 6,300 and 420 for a product of 6,720. Or the student might multiply 100×70 to get 7,000, then 4×70 to get 280, then subtract 280 from 7,000 to get 6,720.

- *Knowledge of place value.* Students can use this information to calculate using large numbers. For example, if the problem is 14×10, an understanding of place value helps you employ an easy equation, $14 \times 1 = 14$, in determining the product of a more difficult equation. Knowing the pattern makes it easy for students to multiply by 100 and 1,000. But students with number sense are able to verify the answer in various ways. They might reason, for example, that ten 10s is 100 and four 10s is 40, and 100 plus 40 is 140.

- *The effects of operations on numbers and the connections between operations.* Number sense enables students to choose the appropriate operation given the situation. For example, a student might recognize that in attempting to calculate $12 \div \frac{3}{4}$, they can think about using repeated addition ($\frac{3}{4} + \frac{3}{4} + \frac{3}{4} + \frac{3}{4} + \frac{3}{4} + \frac{3}{4} + \frac{3}{4} + \frac{3}{4} + \frac{3}{4} + \frac{3}{4} + \frac{3}{4} + \frac{3}{4} + \frac{3}{4} + \frac{3}{4} + \frac{3}{4} + \frac{3}{4}$), or multiplication can help them solve the problem by finding out how many $\frac{3}{4}$s are in 12. A student might reason that two $\frac{3}{4}$s is $1\frac{1}{2}$, so four $\frac{3}{4}$s is 3. There are four 3s in 12, so it takes four times the four $\frac{3}{4}$s, or sixteen $\frac{3}{4}$s, to make 12. Therefore, 12 divided by $\frac{3}{4}$ is 16.

- *How to manipulate numbers in their heads.* Students with number sense are able to use strategies that do not require paper and pencil, their fingers, or a calculator. For example, to compute $1,680 \div 40$, a student might calculate mentally as follows: $1,600 \div 40 = 40$; $80 \div 40 = 2$; $40 + 2 = 42$.

- *How to estimate using benchmark numbers and approximate amounts.* Students with number sense estimate to make sense of quantities. For example, if a student goes to the grocery store with a $20 bill, will they have enough to pay for items that cost

$3.85, $2.95, $1.50, $5.98, $1.99, and $3.05, assuming there is 7 percent sales tax? Students with strong number sense might overestimate the cost of the items, assuming that they cost $4, $3, $2, $6, $2, and $3 for a total of $22 before tax. Calculating 7 percent tax on $20 could be overestimated to be $2, making the $20 insufficient to pay for the groceries.

So how does number sense relate to the basics? One team of educators remarks, "When children have good number sense, they think and reason flexibly with numbers, tend to make sound numerical judgments, and see numbers as useful. When children have poor number sense, they often don't notice when numbers are unreasonable, tend to follow procedures even when another way to reason is easier, and don't have good numerical intuition. . . . Basic facts are also part of number sense. Calculations are easier when we know the basic facts. If our understanding is solid, we can always figure out answers to facts should we forget them, but knowing facts by heart contributes to efficiency and convenience" (Burns and Silbey 2000, 47). This observation is valid regardless of the age or grade of the student. *(See page 186 for information about memorizing the basics.)*

12 Can number sense be taught?

The good news is yes, number sense can be taught. There are a variety of ways to teach number sense. One useful technique is to hold discussions (my friend Karon calls these "number talks") with your students about numerical strategies they use to solve particular problems, and ask them to explain their reasoning. Typically students share a number of strategies for solving the same problem. The discussion continues until no new ideas are offered. This allows students to hear a variety of methods for solving a problem. Explaining their thinking allows students the opportunity to cement their strategy. For example, in the activity *Tell Me All You Can*, students offer as many ideas as they are able about the answers to computation problems without initially finding the answer. While they may know the correct answer, the activity encourages them to compute mentally, estimate, and explore relationships among numbers. One day I asked students to tell me all they could about 3,500 divided by 150. The class was silent so I pressed. "Would the answer be more than ten or less than ten?"

"More than ten, because one hundred and fifty times ten equals one thousand five hundred," Jamila said.

"Did anyone else have the same idea?" I asked. Many students raised their hand.

"It has to be more than twenty," Dee offered, "because if you double one thousand five hundred you get three thousand, which is still too low."

"Does someone have another idea?" I probed.

Debbie said, "It has to be less than twenty-four, because four one hundred and fifties makes six hundred, and if you add six hundred plus three thousand—which is what you'd get with twenty one hundred and fifties—it'd be too much. So the answer has to be less than twenty-four."

Even though the problem I had posed was relatively easy, students had a variety of ways of making sense of the situation and drawing on their number-sense skills.

Another useful way to teach number sense is to link school math to real life. Think about ways to present your class with problem situations that connect to their lives both inside and outside school. For example, for an upcoming school dance, students can be asked to figure the amount of food needed for the dance, determine what admission price to charge in order to cover expenses, figure the quantity of decorations needed, and so forth. Or bring in advertisements from the local paper and have students figure out sale prices for items; for example, if a shirt is on sale for 25 percent off, how much will they pay if the shirt is regularly thirty-five dollars? Having a real context for using numbers is motivating for many students.

Devote time to mental math. Sometimes teachers are concerned that their students are not especially competent calculating mentally, but when asked, the teachers admit that they don't often have their students practice this skill. Like anything else, the more often you ask students to compute mentally and share their strategies, the better they perform the activity. Computing mentally requires students to reason numerically and use methods that are appropriate given the situation, and they may or may not follow standard algorithms or procedures for accomplishing the task. Again, it's important that students share the approach they used, which allows for discussion about efficient strategies.

Don't hesitate to model different ways to calculate. Sometimes students get the message that there is "one right way" when really there are a multitude of ways to get the job done. Granted, some ways may be more efficient than others, but what is important is that students have a way to make sense of quantities that works for them. Students need more than to simply hear the teacher's clear explanations of how to compute. They need many opportunities to work the ideas out for themselves in the context of supportive classroom discourse.

Include estimation in your discussions on number. Much of our daily use of the basics involves estimation. "Do I have enough cash to pay for my groceries?" "About how much money do I need to live on each month when I retire?" "How many skeins of yarn are needed for the scarf I'm knitting?" Because estimation is a skill that develops with time and experience, have students estimate regularly. The more skilled students become at estimating, the better their feel for quantities that are reasonable, and they also develop a sense of relative magnitude. *(See the activity* Multiplication

Puzzlers *on page 126 for an example of how calculators can be used to support estimation sense.)*

Finally, it pays to look for natural opportunities in the curriculum to have students use their number sense. When reading a problem, it is helpful to read it more than one time, for different purposes. While the first read-through is to learn about the learning goals of the problem, read the problem again to think about what basic skills may be embedded. Recognition of basic skills is the first step in finding ways to make those skills explicit. *(See Questions 94 and 95 in Chapter 8, "Using Calculators and Computers," for more about number sense and the basics.)*

13 How do I determine what my incoming students know? What kinds of assessments should I give and when?

When I first began teaching, I was told by my administrator to give my class a "quick" standardized assessment at the beginning of the year to find out what students needed help with. What I found out from administering the multiple-choice test was that the students had a variety of needs: they were all over the board. I had no idea what to do with the information, and I doubted whether a multiple-choice test was a good way to determine their needs, since students could as easily guess and get the right answer as guess and get the wrong answer! The test was not at all helpful in informing my teaching practice. As I reflect on the experience, I find it interesting that no mention was made of learning about students' strengths and using those to build toward their needs—I was just to focus on the students' deficits.

Assessment is an ongoing part of instruction. I've found that the more ways I use to find out what my students know, the more comprehensive picture I have of them as learners of mathematics. One way to determine what they know is to listen to students as they share their ideas during discussions. It can help to have a clipboard with students' names listed so that you can jot down brief observations about the kinds of contributions individuals make. A teacher friend of mine frequently writes her comments on sticky notes that she then dates and files in the students' folders. Another way to assess their learning is to walk around the classroom while students are working and listen to their conversations and explore their thinking by asking questions. Again, if you jot notes to yourself you can save them for future reference. Don't assume that because one student is quick to announce an answer, the entire class understands either the answer or the logic behind it. Don't assume that because students answer as a class, they all understand something. I've learned that it's not very useful to ask, "Does everyone understand?" because a yes or no tells you little. Instead, I make students do something to show me what it is they understand, such as respond in writing to a specific question.

Having students write in math class is a highly useful way to gather information about what they know and understand. Their written responses enable you to gain information about every student in your class. During discussions, you are not likely to hear from everyone, especially from students who are shy about sharing aloud in front of their peers, so asking students to write regularly is useful. *(Chapter 6 gives more in-depth information about writing in math class.)* Incorporate the following kinds of writing prompts into the problems you give:

- Explain and justify your thinking.
- Show your reasoning.
- Solve the problem in at least two different ways.
- I think that the answer is ____. I think this because . . .
- What would happen if you tried this with different numbers (or different shapes or different data)?
- How does this relate to ____?
- Would this work for every problem like this one? Are there problems or situations in which it wouldn't work? Why or why not?

Some teachers give a writing prompt at the beginning of a unit and ask students the same question at the end of the unit. For example, in a unit on area and perimeter, one teacher gave the following assessment at the beginning of her instruction on the topic and again four weeks later:

You want to find the area of this "blob." Your friend has a suggestion: "Lay a string around its perimeter. Arrange the string to form the shape of a square and then figure out its area. The area of the square will be just about the same as the area of the blob." Do you agree or disagree with your friend's method? Or are you unsure? Explain your thinking. Use a sketch if it helps explain your ideas.

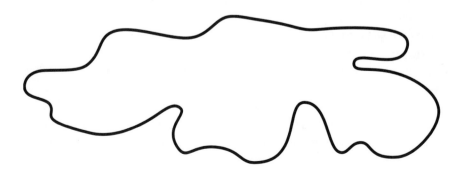

During the second assessment, she returned the papers her students had turned in at the beginning of the unit and asked them to review their answers and explain in writing if they were more firmly convinced of their original response or if their thinking had changed and if so, how.

Don't forget that the assessment materials that come with your instructional materials can also be used as a source of information about what your students know and can do. Often there are suggestions in the materials for when to give each assessment. If not, think about periodically collecting data about students' understanding and proficiency. It's not a good idea to wait until the end of the grading period and then give one major test. Instead, regularly collect information that will continually inform the decisions you make as a teacher.

14 What about standardized testing? What topics are best left until standardized testing is finished? And how can I use testing information from the previous year to learn about my students?

Many important decisions are made about students based on the results of standardized assessments. Such assessments shouldn't serve as the be-all and end-all of your instructional program, but it's important that students are prepared to tackle them. First, find out when students are tested. Then learn what topics are addressed on the tests. Take a close look at your instructional materials to determine how they handle the topics found on the assessment and make a plan for covering those concepts that aren't addressed. Many larger districts that require teachers to use the same instructional materials do this for teachers but in smaller schools and districts, you may have to do this work yourself. Are there topics in the curriculum that are not on the test? If so, it makes sense to teach these topics after testing is over. You'll want to find out about the topics covered, the format of the test, the number of questions, whether or not students are penalized for guessing, how long the test takes, if field-test items are included and how to identify those questions, and so on.

Keep in mind that right answers can hide ignorance! Sometimes students guess the answer correctly. Furthermore, many standardized tests ask only a few questions on each topic, and the format of such tests discourages the inclusion of probing questions. Thus caution is in order when using the test to judge a student's mathematical adeptness. Standardized tests are one measure of a given student's level of achievement at a specific point in time. It's important to use several measures, such as samples of student work, anecdotal records of student participation in class, unit quizzes, tests, and projects, and other open-ended assessments, in developing a broad and comprehensive picture of the student's proficiency.

15 Some of my students really struggle with math. What are some ways to make sure that I meet their needs?

Students who struggle in math class sometimes have difficulties primarily with one or two topics. I've known students who have trouble with number, operations, computation, and algebra, yet are facile in geometry and spatial reasoning, and students whose strengths and struggles are the reverse. I've also taught students who have had difficulties across all topics in math. And of course, many students bump into roadblocks from time to time as new topics are introduced. Regardless of the specific difficulty, there are general responses and interventions that benefit all students. First, at the beginning of the year, set a classroom environment that both supports students and holds them accountable for their learning. This involves setting clear norms for behavior and interaction with one another, having students work with others, and so forth. *(See Chapter 3, "Creating a Productive Classroom Environment," for more information.)*

Another strategy has to do with planning. Select lessons that hold challenges for the entire class while being accessible to struggling students. For example, to help students think about fractions as parts of a whole, you might have them make a fraction kit, which they can then use for several activities and games. In making their fraction kits, students should work with a partner or in a small group. Each person needs several materials:

- five 3-inch-by-18-inch strips of construction paper in five different colors
- scissors
- an envelope to keep their fraction pieces in

Each pair of students (or small group) needs:

- one fraction die, which is a number cube with the following fraction labels: $\frac{1}{2}, \frac{1}{4}, \frac{1}{8}$, $\frac{1}{8}, \frac{1}{16}, \frac{1}{16}$.

Have students take one strip (everyone in the class should choose the same color), fold it in half, and cut it into two equal pieces. Students should label each piece $\frac{1}{2}$, since each is one of two equal pieces. Be sure to talk about the rationale for labeling the pieces $\frac{1}{2}$.

Choose a second color for the second strip; again, everyone should use the same color. Students should fold, cut, and label this strip into fourths. Talk about each piece being one of four equal pieces, or one-fourth. The third strip should be cut into eighths, the fourth strip into sixteenths, and the last strip leave intact and label $\frac{1}{1}$. Explain that it is one of one equal piece, or one whole. Have students turn each fraction piece over and write their initials on the back so pieces can be identified if they get mixed up with others. Students have now made their fraction kits and are ready to use them.

Some teachers wonder if it wouldn't be more efficient to have a teaching assistant or parent volunteer make the kits for students. The point in having students make the kits is that they can see where the fractions come from and how the pieces relate to the fractional notation. Additionally, they can directly compare the fractional parts. Students can see, for example, that $\frac{1}{2}$ is larger than $\frac{1}{8}$ and they can measure to prove that two of the $\frac{1}{4}$ pieces are equivalent to $\frac{1}{2}$. Putting the fraction kit together enables them to develop this knowledge directly.

As with all manipulatives, the value of fraction kits lies in how they are used *(see also Chapter 7, "Using Manipulative Materials")*. There are several activities for which students can use their fraction kit and each supports their learning in different ways. *Cover Up* is a game for two or more players. Each student puts their whole strip ($\frac{1}{1}$) in front of them and they take turns rolling the fraction die described above. What they roll on their turn—the fraction face up on the cube—determines what size piece they can place on their whole strip. The object of the game is to be the first to "cover up" the entire strip. When the game nears the end and only a small piece is needed to win, rolling something bigger won't do—the student must roll exactly what is needed. Students record their moves in two ways. They first record exactly how they covered their strip, for example, $\frac{1}{4} + \frac{1}{4} + \frac{1}{4} + \frac{1}{8} + \frac{1}{8} = \frac{1}{1}$. Then they record by adding fractions that have common denominators, for example, $\frac{3}{4} + \frac{2}{8} = 1$, to help them learn about combining fractions.

The game of *Uncover* helps students with the concept of equivalent fractions. Played with two players or a small group, each player puts their whole strip in front of them and covers it with both $\frac{1}{2}$ pieces. Again, players take turns rolling the fraction die. On their turn they have three options: remove a piece that matches what was rolled, exchange any of the pieces they have for equivalent pieces in the fraction kit, or do nothing and pass the die to the next player. Players may not both remove a piece and trade on the same turn. It's important that students check each other's trades to make sure they are correct.

In these fraction kit games, both the concrete materials and the sequence of the activities support students in making sense of mathematical ideas.

When preparing a lesson, a third strategy is to anticipate the kinds of responses students will give and especially where they'll struggle. Read and think about the notes in your teacher's guide, which can be like consulting with an experienced teacher. You are less likely to be caught off-guard by students' questions as the lesson unfolds. Don't wait until five minutes before class begins to read the teacher notes for the lesson you're planning to teach. Take time to think it through beforehand and envision how the lesson will play out. Think about how you can support students through the potential trouble spots without taking the challenge out of the activity. For example, you might be able to simplify directions, include significant white space on handouts for figuring, or provide organizational supports, such as outlines of tables and graphs that students complete by filling in labels and data.

Don't hesitate to get additional support from other teachers, including your math colleagues and special education and ESL teachers. Meet with them to discuss modifications and accommodations that particular students might benefit from. Finally, talk with parents about options that students can pursue for extra help, such as a math support club or one-on-one tutoring.

16 How do I address the needs of students who finish assignments quickly and who are ready to tackle more complex work?

This goes back to planning and preparation. When you're getting ready to teach a lesson, of course you'll want to have a mathematical goal or goals in mind. It's best to find activities that have the potential to engage students at a variety of levels so that your least-experienced students have access to the mathematics and the most-experienced students are also challenged. While this may sound daunting at first, open-ended problems can offer rich learning possibilities for adolescents of varying levels of expertise. Imagine that students were working on the following problem involving postage stamps:

> *Proper postage for a letter or package sometimes requires you to combine stamps of different values. In this problem, you may use only three-cent and five-cent stamps, but you can use as many of each as you want. What postage amounts would be impossible to make? Is there a largest impossible amount? Explain your answers.*

Students who need a bit more challenge might investigate the following after they'd explored the initial question:

> *Do the same investigation, but this time you are limited to eight-cent and five-cent stamps. Again, you can use as many of each as you want.*

Finally, students might attempt to generalize for any two-value combination of stamps:

> *Continue with other two-value combinations. For each, find out how many impossible amounts there are, and if there is a largest impossible amount (and what it is if there is one).*

Or, say you taught students how to play the game *Poison,* which involves logical reasoning. In the game, two people play against one another. They need thirteen objects. Players take turns removing either one object or two objects from the collection at a time. They play until all objects have been taken. The last object to be removed is considered to be the "poison." Whoever gets stuck taking it loses. While everyone in class can play the game, students should be encouraged to search for a winning strategy, because there is a way to win the game every time. To push students to generalize from the specific case to the general, you can offer interested students the following extensions:

- Change the number of objects in the collection to, say, fourteen or fifteen objects. Can you still find a winning strategy?
- Change the number of objects you can remove on each turn. What if you can take one, two, three, or some other number of objects?
- Can you find a master strategy so that regardless of the number of objects and how many are taken on a turn you can win every time?
- Change the name to *Dessert* and play the game so that the person who takes the last object gets the "dessert" and wins. How does this affect your strategies?

The idea here is knowing how to nudge students toward more complex thinking, such as generalizing their specific strategies, so that you can plan questions that will engage students who need an additional challenge. As you gain experience you'll get better at pulling these questions out of your proverbial back pocket, but it always helps to plan them in advance.

17 How do I encourage my students to become independent learners? It seems as though the moment I finish giving directions, half the hands in the room go up.

This is a common situation for middle school teachers. Adolescence can seem to work against individual initiative at times! Having clear and consistent classroom guidelines or expectations about what students should do when they need help is therefore essential.

My middle school students sat in groups of four. I explained to the class at the beginning of the year that if they had questions related to a lesson activity, they were to first check with their table group to see if anyone could help them. If nobody at their table could answer the question, they were all to raise their hands, which would be my signal to check in with them. Sounds good in theory, right? I remember the first time I explained this to the class and then proceeded to give directions about a task. Want to guess what happened? The usual flurry of hands went up as soon as I was done. They seemed to be testing me, to see if I really meant what I'd said. I approached the nearest student whose hand was in the air. "Sam, looks like you've got a question," I said. "Have you checked with the others in your group to see if they can help you?" "Uh, yeah," he replied, and before he could go any further I turned to one of his groupmates and asked, "Bianca, what's Sam's question?" She gave me the *I have no idea* look and I said evenly to Sam, "Let me know when everyone at your table has the same question," and I left their table. Sam looked rather stunned that I wouldn't answer his question—indeed, I didn't even give him the chance to ask it—but he quickly learned that I meant what I said, and the rest of his group learned the same. This scenario was repeated at five different tables that period. But it didn't take long for students to get in the habit of checking in with one another. I sometimes find that students' questions are along the lines of "Where do I find that information?" or "What do we do when we're done?" and these logistical kinds of questions are easily answered by others in the group. I figure that in a group of four, there's at least one person listening to me at any time, so they are generally in good shape to support one another. When students' questions are about the mathematics, they have others at their table to support them.

If your students don't sit in groups, you can ask them to check with two or three others who sit nearby before asking you.

18 What are effective ways to structure my daily math period?

The answer depends on many factors: for example, how much time you have to teach math each day, the instructional materials you plan to use, how you will organize students to work—in small groups or individually—and on the lesson you want to teach. There's no one best way, obviously. Routines are useful, yet varying the structure of your math class helps maintain interest and motivation.

If you teach in a setting where you see several different classes over the course of a day, using a routine such as a warm-up at the beginning of class can help students transition to your class. A warm-up can take more time than intended, so if you have short periods—forty-five minutes or less—it's a good idea to connect the warm-up to the lesson you're planning on teaching. *(See Question 19 for more information on warm-ups.)*

Teachers typically introduce student problems or activities that the class is expected to work on. Keep such introductions short and share only enough information for students to get started. A common mistake that inexperienced teachers make is to give away too much of the mathematics so that there's not a lot for students to think about as they work on the problem.

Students then need time to work on the task. If they're working independently, you probably won't encourage them to talk with one another as they do the activity. But if students are organized in groups or with a partner, it's important that they're clear about your expectations for the group as well as for themselves as individuals. *(See Chapter 3 for more about having students work together in groups.)* It can be useful to tell students how much time they have to work on the activity before you call them together for a discussion. Some teachers use a kitchen timer to help them stick to their time limits.

Finally, discussing the task or problem is essential. This is when students share their approaches, strategies, generalizations, and talk about what they did when they got stuck. It's a time for the big mathematical ideas to become public. *(Chapter 5, "Leading Class Discussions," provides more information.)* Discussions are essential—always make time for them, even if this means holding them the next day.

Some teachers structure their math class by occasionally providing a collection of a small number of activities (five to seven) that students work on for a week or more. These activities have been carefully chosen to address specific mathematical goals. Described by some teachers as a "menu," they allow students choice and control over their learning and engage them in tasks that simultaneously support, motivate, and challenge them.

Imagine that you've walked into a typical middle school math classroom where students are working on a collection of independent activities. You notice first that many students are working and talking with one another—some in small groups of three or four and others in pairs—and yet other students are working independently. There is noise, but it is productive noise. Everyone seems to be working on something different, but when you look more closely you see that there are only four or five different tasks and that they all deal with the same topic but in different ways. The teacher is not at the front of the room lecturing. She is standing next to a table where three students are working, and a lively discussion between the four is in progress. All students are engaged, on task, and busy. Each student is involved in doing something that is important. There is occasional movement as students retrieve supplies from a central location, borrow and return manipulative materials, and turn in work. As you survey the class you get a sense of purposefulness. After a few minutes the teacher calls for the students' attention and reminds the class that tomorrow they will discuss one of the activities and to be sure that they've completed it by then. (A Collection of Math Lessons from Grades 6 Through 8 *[Burns and Humphreys 1990] has more information about using "menus" with adolescents.)*

However you choose to organize your math period, be sure to clearly communicate to students your expectations and guidelines for how to handle classroom transitions. Your expectations should address behavior during lessons and class discussions, how to get and put away materials such as graphing calculators or manipulatives *(see Chapter 7)*, parameters for where they should sit, turn in work, and so forth. The clearer you are, the less time you will need to spend managing students as the year progresses and the more time you will have to learn about them and their ways of thinking about mathematics.

19 My colleague has his students complete a warm-up at the beginning of class. Are warm-ups an effective teaching tool?

Warm-ups are short activities that are typically done at the beginning of math class and are not intended to take a lot of time. They can be useful in providing a transition to class for both students and teacher. For students, a warm-up activity helps them make the mental shift from their previous class to math class. Teachers can use warm-up time to make the final preparations for the day's lesson and to take care of logistical matters such as taking attendance and putting away papers from a previous class. Sometimes warm-up questions are written on the board or overhead projector; at other times, the warm-up may be photocopied and distributed to students when they arrive.

Mathematically, a warm-up might launch the day's lesson or it might extend something students did the previous day. A warm-up can be used to provide students opportunities to enhance and maintain their number and operations sense as well as their computational fluency. Additionally, many teachers use one or two standardized test questions as a warm-up. Despite their benefits, some teachers may not have enough time in their math period to devote to a warm-up. If you have only forty minutes or so, you may decide that you need the entire time for the lesson.

To use warm-ups effectively, it is helpful to think first about their purposes for students. Then, establishing a routine and communicating it clearly to students is necessary if they are to get the most mileage out of the warm-up. Finally, it helps to be aware of the pitfalls associated with using warm-ups.

Students need to know what you expect from them during the warm-up routine. Explicit teaching of your expectations is crucial if students are to successfully meet them. It will help to think through the following questions: Are students to be in their seats during the warm-up? Silent? Or is conversation and movement around the classroom okay? Many teachers insist that for warm-ups to be effective, students must work individually and silently for several minutes.

How will you process the warm-up? After students have had a couple of minutes to work on the warm-up problem or problems, many teachers initiate a discussion that focuses on how students figured out the answer. Have students explain their reasoning to the class. It is not sufficient to just write the answers on the overhead. Working on the problem, *then* explaining and listening to others' solution methods supports students' learning.

What do you expect students to be doing when someone is addressing the class? Do you expect them to take notes on notebook paper or on a warm-up sheet distributed to students? What should they do with it when they are finished? Do students keep their warm-ups in their binder? Portfolio? Will students be expected to use the warm-up again later in the year, such as before testing? Finally, think about how you will manage the transition to the lesson with a minimum of disruption.

Potential pitfalls of using warm-ups include the following:

- *Neglecting to establish a warm-up routine.* Without a routine for completing warm-ups, students are likely to turn this activity into a free-for-all. Management problems are more likely to occur when students cannot anticipate what should happen during the warm-up, and instructional time can easily be wasted.

- *Not valuing the warm-up as a learning tool.* If you don't take the learning potential of a warm-up seriously, neither will your students. The warm-up should be as well thought out as your lesson plan.

- *Not following through with or inconsistently enforcing expectations for students (or worse, having no expectations).* Regardless of how you choose to use warm-ups with your class, students must be explicitly taught how they are to handle the warm-up routine. At the beginning of the year, many teachers explain the warm-up routine to students and model the successful completion of a warm-up. Attention must be paid at this point to how students follow through with your instruction, and corrections made as needed until your expectations are internalized by students and the routine is firmly established.

- *Spending too much time on the warm-up.* If the warm-up is too long, time is taken away from the lesson for the day. It's easy to fall behind schedule, jeopardizing students' potential mathematical achievement as well as putting next year's teacher (and students!) in a bind because you didn't stay on schedule.

- *Not being clear about the purpose of the warm-up.* It's important to explain to students, administrators, and families the purpose of doing warm-ups. Providing clear, specific information helps everyone support your program.

- *Providing too much time for class discussion of a warm-up.* Sometimes teachers allot too much class time to discussing the warm-up. The same kinds of issues that arise when a warm-up is too long apply in this situation. Some teachers ask a responsible student in class to monitor how much time is taken for each part of the warm-up, including the time taken to work on the warm-up as well as to discuss it. This information helps the teacher become more aware of how time is spent during the period. Generally speaking, a warm-up should take about five to seven minutes for a forty-two-minute class, and eight to ten minutes for a longer period.

Before beginning warm-ups with your class, you need to decide whether the warm-up work will be scored or graded. Some teachers prefer not to grade warm-ups. If you choose to mark them, what criteria will you use? Some teachers score students on completeness of the task. Others give extra credit or a higher grade if students take notes on the warm-up during the class discussion, for future reference when preparing for a test, for example. Will you deduct points if a student forgets to sign their name? Will you consider legibility? Completeness? It's important to think through how you will score or grade warm-ups, and clearly communicate with students and families the criteria you use in marking warm-ups.

	Creating a
Three	Productive
	Classroom
	Environment

"I'm really frustrated with my students," David confided to his colleague Lauren. "They don't follow directions, they're constantly moving around the room without my permission, and they seem to be focused on everything except mathematics! They also put each other down . . . and nothing seems to work to get them to behave decently. I just don't know what to do."

"I know what you mean," Lauren said with a sigh. "My students behave okay for the most part during whole-class lessons, but when it comes to working together in small groups, everything falls apart. Either one student ends up doing all of the work for the group while the others slack off, or nobody takes charge and the group gets nothing done. I've just about given up on groups."

Comments like these are frequently voiced among middle-grades math teachers. Effective classroom-management skills, a positive learning environment, and attention to group dynamics are essential ingredients in math learning because little learning goes on in the absence of a supportive classroom environment. It is difficult at best for teachers to teach effectively and students to learn without all three in place. In an effective math classroom that supports significant, long-lasting math learning, the culture is one in which collaboration, inquiry, mistakes, and disequilibrium all provide opportunities for learning.

In order to create such a culture, you need to learn about your students. It's important to know about adolescents generally—their interests as well as what they are capable of and what is challenging for and about them. For example, fitting in (or, sometimes, standing out) is important to adolescents. Saving face can be critical. Their sometimes wacky behavior can often be attributed to normal developmental changes. This doesn't excuse bad behavior, of course, but can certainly help you understand the behavior and develop appropriate intervention strategies. Adolescents need structure, consistency, high expectations, encouragement, and respect. But in addition to this general knowledge about adolescents, you also have to get to know your students as individuals. Who is on an Individualized Education Program (IEP)? Which are English-language learners?

Gifted? Struggle socially? Need help with organization? Are independent learners? Unwilling to take risks? Are dealing with difficult issues outside of class? You especially need to be mindful of the cultural, class, gender, and language differences among the individuals who make up your class. All of these ways of knowing students makes it possible for you to create a classroom environment that promotes learning.

20 What are the benefits of classrooms where students are encouraged to work together to solve math problems and discuss their solution methods?

Working together supports students in the development of mathematical ideas. Students who work together effectively typically produce work that is more thorough and complex than what they would be able to do on their own. A group that works well together contributes to the development of individual members' self-assurance. Students who regularly check their ideas out with one another and get feedback about their thinking are more likely to develop confidence about their ability to do mathematics.

Students tend to learn more when they work together than when they work alone. Talking about mathematics facilitates learning. When you explain something to a partner or group, it helps you clarify and cement your thinking. When you listen to others, their ideas can help challenge your thinking, providing perspectives and strategies that you might not have considered otherwise. Working with a partner or in a small group provides opportunities to build knowledge together and learn from one another.

21 How do I set up a learning environment in which students are willing to take risks?

Think through the classroom environment and the rationale for all of your decisions. Do so through the eyes of your students as well as through your "teacher lens." It's the messages that the learning environment is sending out, both explicit and implicit, that will influence whether or not students become academic risk takers. Continual attention to these messages is necessary; your approach requires consistency and follow-through. When you speak to your students, do so respectfully, even if you are feeling angry or annoyed with them. It's seldom necessary or productive to yell at students. One teacher remarked to me that she noticed a positive change in her students when she altered how she interacted with them. When she noticed a student off task, for example, she used to say loudly from across the room, "Doug, get to work!" Now she kneels next to the student and says quietly, "Looks like you're having trouble getting started." She said that she and her students are happier and more productive since she's made this small change in her management style. Sometimes it's helpful to share with your students your rationale for the decisions you make. This sends the message that you have carefully considered your decisions, and sharing your reasons with students conveys respect. Asking for students' opinions and ideas also tells them that they are valued members of the classroom.

When students act out, it is usually because their intellectual and/or emotional needs are not being met. Teachers who have taken the time to get to know their students are in the best position to address such behavior. Survey data repeatedly reveals that adolescents' favorite teachers are those who are "strict" yet kind; students tend not to act out in their classes. These teachers set high expectations and are effective classroom managers; they consistently monitor students' behavior. The tasks they assign are both appropriate and meaningful.

A colleague of mine once shared with me her struggles with classroom management. Her students were disrespectful when fellow students addressed the class, and seldom completed class work. My colleague knew it was time to "start over" with the class, even though it was midway through the first quarter. She decided to collect some data from her students in an effort to involve them in what was wrong and, hopefully, contribute to a solution.

"I have an anonymous survey for you to think about and then fill out," she told her students. "'Anonymous' means that you don't need to put your name on the paper. There are three questions for you to think about and then respond to," she continued. "May I have a volunteer to read the first question aloud?" Her survey looked like this:

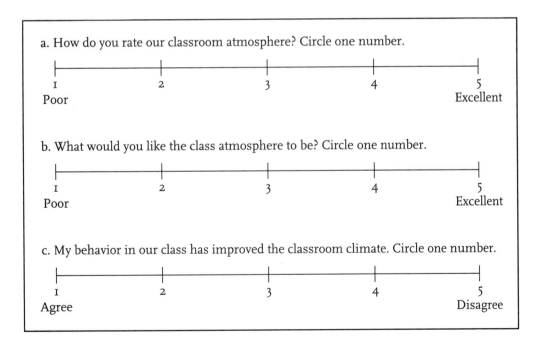

a. How do you rate our classroom atmosphere? Circle one number.

| 1 | 2 | 3 | 4 | 5 |
| Poor | | | | Excellent |

b. What would you like the class atmosphere to be? Circle one number.

| 1 | 2 | 3 | 4 | 5 |
| Poor | | | | Excellent |

c. My behavior in our class has improved the classroom climate. Circle one number.

| 1 | 2 | 3 | 4 | 5 |
| Agree | | | | Disagree |

After a student read each question aloud, the teacher led a discussion about what the questions meant and how the scale worked, and then each student filled out their survey and turned it in. The teacher compiled the data shown on page 33.

The next day, the teacher asked her students to predict the results and then she put the completed graphs on the overhead one at a time, using the data to initiate a discussion about possible changes that they could make in their classroom as a result. The teacher took a risk when she acknowledged to her class that things weren't going well. Initiating both the survey and discussion encouraged her students to risk discussing and acknowledging their behavior.

22 My colleague suggested that I talk with my students at the beginning of the year about working productively together. How can I do this without lecturing my students about the importance of cooperation?

Initiate a class discussion about working together. Adolescents appreciate having their opinions considered. Furthermore, talking about working together conveys the message that working together is important!

Explain to your students that sometimes they will work by themselves and at other times they'll work with a partner or in a small group. Ask students what they like and dislike about working alone, working with partners, and working in a small group. Typical responses on working alone might include, on the positive side:

I get to organize the work the way I want to.

I don't have to worry about disagreeing with a partner.

I have the time I need to think about the problem without being pressured by a partner.

Negative comments might include:

There's nobody to ask for help.

Sometimes I get stuck.

I worry about finishing too slowly.

When it comes to working with others, students often report:

I get more ideas from the group.

If one person doesn't know the answer to a question, someone else in the group usually does.

Your ideas grow when you put them together with other people's ideas.

a. How do you rate our classroom atmosphere?

```
                    X
                    X
    X               X
    X               X           X
    X               X           X
    X               X           X
    X               X           X
    X               X           X
    X               X           X           X
    X               X           X           X
  ──┼───────────────┼───────────┼───────────┼───────────────────┼──
    1               2           3           4                   5
  Poor                                                      Excellent
```

b. What would you like the class atmosphere to be?

```
                                                                X
                                                                X
                                                                X
                                                                X
                                                                X
                                                                X
                                                                X
                                                                X
                                                X               X
                                                X               X
                                                X               X
                                                X               X
                                                X               X
                                                X               X
                                                X               X
                                                X               X
                                                X               X
  ┌─────────────────────────────────────────────────────────────
  1               2           3           4                   5
  Poor                                                      Excellent
```

c. My behavior in our class has improved the classroom climate.

```
                                                X
                                                X
                                                X
                                                X
                                X               X
                                X               X
    X                           X               X           X
    X               X           X               X           X
    X               X           X               X           X
  ──┼───────────────┼───────────┼───────────────┼───────────┼──
    1               2           3               4           5
  Agree                                                  Disagree
```

Negative observations about groups might be:

One person gets stuck doing all the work.

Sometimes people goof off and the work doesn't get done.

Someone takes over and you're just sitting there.

Sometimes group members are mean or make you feel dumb. *(See Question 26 for more about status issues in groups.)*

After students have shared their thoughts about various group configurations, explain that sometimes it's helpful for them to work alone on a problem, especially when you are trying to learn what they know and can do independently, and that this year there will be times when they are expected to work by themselves. Tell them that at other times, it's helpful for them to work with a partner or small group. When they work with others, their thinking is enriched by hearing different points of view. They gain the opportunity to work out what they know by explaining it to someone else. Furthermore, as they challenge others' mathematical ideas, they're engaged in high-level thinking. Tell the class that what's important when they work with others is that they listen respectfully and carefully to one another and that they make sure that everyone in the group is involved in the task. Acknowledge that this is not always easy, but that it is important and that you as the teacher will help ensure that the groups work together effectively.

At the beginning of the year, Joan asks her students to individually write the answer to three questions before they begin working together in groups:

When you work with a group to solve a mathematics problem,

1. What do you want others to know about you as a learner of mathematics?
2. What supports your learning?
3. What is it that you do not want others to say or do?

She then asks students to take turns and share their answer to question number one with the members of their group. Next, Joan asks each student to share their responses to the second question with the members of their group. She then asks a representative from each group to report to the whole class something that came up in their group for question two, while Joan writes their ideas on a poster. Group representatives are selected in a way that ensures equal and random participation. For example, Joan might say, "The first time we go around the room and hear from each group, I'd like the person whose birthday is closest to today to report." Typically, the poster looks something like this:

What makes groups work well
• Everyone takes turns to speak.
• When you get confused, others can help you.

- You can help by asking a question, not just by telling the answer.
- Everyone is patient with one another.
- You listen to different ways to solve the problem.
- Everyone does their share of the work.
- Everyone has time to think before your group talks about the problem.
- Even if you disagree you listen to what others have to say.
- You get help when you ask for help.

Next, Joan asks her students to report to the class what gets in the way of their learning in a group; that is, what they don't want to hear or see. The following is a representative list:

What we don't want to see/hear when we work together
- put-downs
- someone taking over the work
- people who give the answer before you've had time to think about it
- group members who slack off and don't do their share
- eye rolling
- when someone in your group is off task

Typically, students have no problem generating both of these lists. But creating the lists doesn't mean that students will automatically behave appropriately. After the lists are complete, Joan posts them prominently in the classroom as a reminder of the ways they can maximize their learning when working with others. She frequently refers back to the lists after the class has finished a group task. "What helped your group get the job done today?" she might ask. As students report, Joan points out the appropriate item on the first chart. If something comes up that is not on the list, she asks students if it should be added to the chart. The list thus becomes a "living document." By the same token, it's helpful to discuss when things *didn't* go well in a group, referring to the second list.

The process of creating the lists is as important as what ends up on them. Sometimes teachers complain that writing lists like these is not a good use of precious class time. Teachers wonder if it wouldn't be more efficient to just post teacher-created norms and tell the class to abide by them. I agree that this list-making process does take a bit of time. But the payoff is well worth it. By taking the time to have students generate the ideas, you send the clear message that their ideas are important and that working together productively is something you expect them to do and get better at as the year progresses. Furthermore, students have more "buy in" to working in groups when they know that there are expectations for behavior. Honest, tactful, and regular discussions are necessary to ensure smooth functioning of groups, and this type of activity is a good way to get the process started.

23 How can I support students in working together productively?

Here are some tips for helping students work together successfully. Make sure that you've chosen a task that is appropriate for group interaction. If you give students a task that is easily accomplished individually, there is little incentive for groups to work collectively on it. If the task you've chosen is inappropriate for group work, you'll know it almost immediately—because students will grumble, act out, and generally create chaos in the classroom.

It's advisable to provide students time initially to work on the problem alone and silently. Explain why this process is valuable. The effectiveness of the group depends on the quality of individual contributions. Individuals must have the chance to think about the problem on their own, without being influenced by others. They need to know their own thinking about a problem before they can voice their thinking in front of a group. Moreover, students who are less confident will be more likely to let their partner or group mates do the thinking for them if they are not encouraged to spend time with the problem on their own.

After a few minutes have passed and you sense that students are ready to interact with one another, tell them to take turns *(consider using a protocol—see Question 24)* and share their thinking with a partner. It helps if the partner is sitting nearby so that students do not need to leave their seat. You might use a timer to ensure that each person has an equal amount of time to share their thinking. After sharing their initial thinking, ask students to come to a consensus about their approaches and answer. This process will likely take some time. Before they do this for the first time, discuss and consider role-playing what appropriate sharing looks and sounds like. The more clearly you communicate what it is you expect, the more likely students will meet or exceed your expectations.

24 What tools or protocols support equitable participation in the math classroom?

One simple tool or protocol is a dyad. In a dyad, two students pair up. Each has the same amount of time to respond to a question that you want them to think aloud about, such as *"What are all the ways that a variable can be used in math?"* or *"What was challenging for you about solving this problem and how did you work through the challenge?"* Each student then has a specified amount of time, such as two minutes, to respond aloud to the prompt while their partner listens attentively without interrupting. At the end of the allotted time (it helps to set a kitchen timer), the second person has their two minutes to talk. They are not responding to what the first person said, but voicing their own thinking about the question. Sometimes in groups, the most dominant student or the one with the most social status takes over and other students have difficulty getting a word in edgewise. Dyads give all students the chance to articulate their thinking in front of an audience; dyad sharing can serve as a useful prewriting activity.

Another tool that helps each student have an equal voice is the *Go Around One* protocol. In this protocol, person A reports one idea to their group. While person A reports, the other group members listen carefully without interrupting with either questions or comments, and without giving nonverbal cues to indicate that they agree or disagree. When person A finishes reporting their idea, person B shares a single idea with the group. Again, group members listen attentively as described above. The process continues until all group members have reported their idea. Then the group discusses the ideas, sharing the airtime equally.

In the *Go Around All* protocol, the process is the same but instead of reporting only one idea, each person in turn reports all of their ideas. The *Go Around Timed* protocol is similar to *Go Around All* except each student reports until the teacher rings a chime or other signal and says, "finish your sentence and rotate to the next person." At that time, the person finishes their idea and the next person reports until time is called. The process is repeated until every student has had a chance to report. In the *Go Around Timed* protocol, it helps to have groups of equal size. If not, the group with a different number of participants must adjust their timing.

25 Sometimes when students work together, one or two individuals do most of the work while the other members goof off. What are ways to ensure individual accountability when my students are working with a partner or group?

This is a common problem because of the status that adolescents typically assign their peers. *(See Question 26 for more about addressing status issues in groups.)* Long before they ever reach your classroom, students have formed opinions about who is "smart" or popular, and what the risks (as well as benefits) are in working with different students. Again, students benefit from expectations that are both clearly understood and consistently enforced by both teacher and students. Many teachers report benefits from assigning roles, such as team captain, facilitator, recorder/reporter, and resource monitor. Likewise, using math "task cards," in which expectations for each role are explained in writing, is helpful. *(For further information, see* Designing Group-work *[Cohen 1994]. It describes the benefits of roles, group-worthy tasks, and intervention strategies in detail.)*

26 Students' contributions are sometimes dismissed or marginalized by those in their small group who are perceived as being "smarter" or more popular. How can I address this?

Addressing status issues is an important job for a math teacher. First, you have to be able to recognize when status problems are affecting a group. Observe your students working together. Do you see some students hanging back or not participating? Are others dominating? These are often signs that "agreed upon" social rankings are operative within a group, based either on perceived academic ability or popularity. One powerful way to address status bias is by helping students recognize that all of them together are much "smarter" than any single one of them alone. Often many kinds of skills and knowledge are necessary to solve problems. Remind students that no one student has all of the abilities that are needed, but each person has some of them and that their group collectively will have all of them.

Johnny had prepared a pattern problem using tiles for his students to work on in groups. Before he gave the problem to the class he said, "For this task, you need to be able to see and describe patterns, connect a visual pattern to numbers, and represent the pattern using symbols. You'll also need to analyze information and make generalizations. You'll represent information in various ways and make connections between and among them. Communicating your ideas in writing is also important." He reminded the class that nobody in the room had all of those abilities but everyone in their group had some of them and that their group collectively would be able to solve the problem with the skills they possessed as a whole.

To encourage students to see value in their differences, the next step is to listen to groups as they are working and write down what you hear or see that both furthers the mathematical understanding for the group and moves the task along. It's essential to record who makes each contribution. These offerings then need to be shared explicitly with the whole class (e.g., "Li noted that the numbers grow in a predictable pattern, which helped her group notice . . ."; "Carmen made a chart with two columns, helping her group organize the information so that . . ."). "Status" is thus assigned to the individuals whose contributions enabled the group to work together to solve the problem.

27 When my students work in groups, I sometimes hear them putting each other down. What should I do when this happens?

Learning is an emotional experience. Recall the joy on a student's face when understanding dawns and he shouts out, "I get it!" At the same time, it can be disheartening for students when they make mistakes or struggle with understanding. Think about a time when you heard a student make a negative comment to another, who then shuts down. It is critical that put-downs are addressed by either you or your students every time they are heard. Otherwise, students get the clear message that your classroom is not a safe place for them to take academic risks, which is likely to diminish their learning.

When you hear a put-down, the first thing you have to do is decide how serious it is—which will in turn dictate how it should be handled. If it is less serious, such as when a student says, "Your writing is messy, let me do it," watch the students involved to see what happens. Does the "victim" address the situation or ignore it? If the situation is handled well, great! If the situation is not handled well, it will require your intervention.

Next take both (or all involved) students aside to hear all sides of the story. Adolescents are highly conscious of saving face, so having this conversation privately, out of earshot of others, is important. You might start by asking the offending students why they think you called them aside, keeping your tone neutral. This gives students a chance to own up to their mistake and deal with it. If they refuse to take responsibility, explain calmly to the offending students that their comment or action is a put-down and is not acceptable in your classroom, before deciding what consequences, if any, are appropriate. And remember, you don't need to determine a consequence immediately in most situations. *(Jim Fay and David Funk's* Teaching with Love and Logic *(1995) has more information about classroom management and discipline.)*

28 My students can't handle working together. Why would I even want to consider letting them sit in groups?

Let's address the negative expectation in this question first. If you assume that students can't handle working together, they won't be able to. Productive group work involves more than teacher expectations for success. Students must be *taught* how to work together with others.

We are creatures of habit, and many teachers tend to teach their students in the ways they were taught, right down to desk arrangements. There are few professions, however, that do not require their employees to work together. Teaching students how to work together is therefore vital.

It can be helpful to record your students' growth in this area to remind you and your students later in the year how far the class has come. Some teachers videotape their students working together early in the year and again later in the year to compare the behavior and celebrate students' efforts. Other teachers keep notes for themselves for the same purpose. Keep in mind that it can take a long time for students to internalize your expectations. Sometimes it isn't until midyear that I feel like my students are behaving the way I expect them to!

Finally, talk with your colleagues whose students are also working together in groups, to share ideas and tips. Get support for yourself; don't try to go it alone.

29 I'm reluctant to have students come to the front of the class to share their thinking because I'm concerned that other students will be disrespectful. Is it really worthwhile having students present their thinking in front of the class?

If you don't let students come to the overhead or the board to present their thinking, you become the "sage on the stage," creating a passive classroom. Students need to take active control of their learning, and they should have the responsibility of sharing what they have learned publicly with the class. This has several benefits. Students who present their thinking in front of an audience are given the opportunity to more firmly convince themselves about what they know or have questions about, and those listening to the explanations benefit from hearing a variety of ways of thinking about a problem. Both are essential ingredients for learning. Students learn that you consider their ideas to be important, which is often motivating. And many students take more care with their work when they know they will make it public.

That said, it is extremely important to make sharing their thinking with the class a safe proposition for students. Thus you have to teach them what you expect from both the presenter and the audience and then follow through consistently and respectfully until students have internalized your expectations. (*See Chapter 5, "Leading Class Discussions," for more information.*)

30 I'm thinking about how to physically set up my classroom. I thought about putting the desks in rows. Another teacher suggested that I move the desks to form small groups. Is this organization recommended?

Productive interaction between students supports learning. Arranging desks into small groups facilitates discussion. When students have questions, they have access to a group of peers who can help them. Keep in mind that everything about your classroom communicates about what you believe is important. Arranging desks into groups gives the message, "Talking productively with one another to support your learning is so important, I am going to arrange your desks into groups to make sure that it happens."

31 Should I let my students sit where they want?

There are pros and cons to allowing students to sit where they wish. When students choose their seats, you are able to see who gets along well with whom, and where potential problems might crop up. Allowing students to sit next to their friends presents drawbacks, however. For some students, the temptation to socialize outweighs the desire to focus on math. Students may rely on established power structures within the relationships and more dominant students may take over the work. Furthermore, students who always sit with the same students hear only how those particular individuals think.

32 What about assigning seats?

Like allowing students to choose their own seats, assigning seats has advantages and disadvantages. First the advantages. Assigning seats allows the teacher to control who has access to conversation with whom, which means that students who may not typically interact have the opportunity to do so. It also allows you to separate students who you know have difficulty working together. Now for the drawbacks. Assigning seats can be time consuming. Some teachers assign seats so that "difficult" children are seated near "quiet" or "obedient" students, a strategy that can lead to resentment on the part of the latter. Other teachers arrange their seating charts so that in a group of four working together, there is a "high-achieving" student in the group, two "average" achievers, and one "low" achiever. More often than not, adolescents also have an intuitive sense of why they are placed in a particular group or near a certain student. With their heightened sensitivity to fairness, this can create problems and cause students to internalize negative messages about their perceived abilities.

If your students sit at tables or at individual desks clustered together to support group work, there is an effective way to handle seating students so that seats are assigned randomly. Here's how it works: ahead of time, make a number label for each of your tables. These labels can be made by folding two five-inch-by-eight-inch index cards in half, then stapling the edges together to form a four-sided box without a lid or bottom.

Using a marker, write a *1* on the four sides of one of the table labels, a *2* on the second label, and so on until you've made as many labels as you have tables. Place one label in the middle of each table. Next, take a deck of playing cards and pull out as many 1s as you have seats at table one, as many 2s as you have seats at table two, and so on. Shuffle the cards.

When students arrive in class, hold the cards face down and fan them out. As each student enters, have them pull one card from the deck you are holding. The number on the card that they pick tells them which table to sit at. Students should take their playing card with them to the table that matches the number on the card and drop the playing card inside the label box, making collecting the cards easy for you. It's important that students don't trade playing cards with one another, so you'll need to watch them carefully the first few times they get new seats. I always make students show me the card that they picked before they head for their new table. I can't remember what card each student got, but I can remember those four or five students who I'm most concerned might trade cards or otherwise find their way to a different seat.

Here's another way to assign seats that works regardless of whether your students sit at individual desks in rows or at tables or clusters of desks. Ahead of time, write the name of each student on an individual craft stick or small slip of paper and put them in a sack or other container. Have students gather their belongings and stand around the perimeter of the room. Begin by walking to any seat at random. Sit down at that seat and tell the class, "The person whose name I draw will sit here." Draw a name from the container and announce whose name was randomly picked. That student should take their belongings to their new seat and sit down. Ask the seated student where the next person should sit and have the student clearly select one seat by pointing to the seat. Then the student should draw the second name out of the container. This determines who will sit in the seat selected. Repeat the process until every student has a new seat.

So instead of selecting a person and then determining where they will sit, the seat is chosen first and the student who sits there is randomly determined. Thus the class knows in advance where the next person will sit, but not who will sit there. This process provides choice and control, which is motivating for adolescents. And students see that you clearly are not "rigging" the seating chart. A drawback to this system is that it is somewhat time consuming and can be complicated to explain to students initially.

33 How often should I have my students change seats?

About every two to three weeks seems to work well for many teachers. The idea is to keep students together long enough so that they learn how to work with their group mates and get to know each other's strengths and weaknesses. Changing seats more frequently doesn't encourage students to persevere when the group dynamic proves challenging to them—kids don't have an incentive to work through their difficulties when they know they'll soon be switched to another location and other partners. It's important that they feel some sense of investment in working through interpersonal issues.

34 What sorts of things might I post on the bulletin boards in my classroom?

Student work, class charts, students' posters, work in progress . . . the idea is for students to see themselves and their work reflected on the walls of the classroom. This provides students with an incentive to produce high-quality work and again communicates to them what it is that you value.

It's very helpful for students to see posted a list of what they are learning in class. Some teachers post a student-friendly list of standards to help communicate to students and visitors what math students are learning. Examples of student work that meets or exceeds the standards set by your school or district should also be posted to help communicate to students what kind of work you expect them to do in your class. Post your classroom expectations as well. A student-made poster outlining expectations is particularly nice. Many teachers post "word walls," or vocabulary lists with illustrations and examples from the unit currently being studied.

Four | *The First Week of School*

Johanna had been teaching math for many years when Spencer, a new teacher, moved in to the classroom next door and asked for some help in setting up his room. Johanna readily agreed. They pushed the tables together to form groups of four, set up bulletin boards with space for student work, and decided on the best places to put the manipulatives and calculators.

"So how are you feeling about the beginning of school?" Johanna asked.

"I'm concerned about inundating my students with rules and regulations on the first day," Spencer explained. "Last year when I was student teaching, my mentor teacher told me that he thought it was important to tell students everything they needed to know about his expectations for the year on day one. But you should have seen the glazed eyes in the classroom. . . . The kids seemed overwhelmed with all the information and, frankly, bored. And at the end of class, many complained about how hard it was to sit still all day after being on summer vacation! I think explaining your expectations is important, but I want to do so in a way that engages students rather than puts them off. And isn't it a good idea to get students excited about math right away?"

"I understand your concerns," Johanna replied. "And you've got good instincts about getting the new school year off on the right foot. I remember being nervous before my first day of teaching that students would defy me and I wouldn't be able to control the class."

"I've worried about that too!" Spencer said. "I'd love to hear about how you get the year started. I have a few ideas but I'm nervous and want to get it right."

Spencer's concerns are legitimate and genuine. Furthermore, they are common to both new and experienced teachers. The first week of school sets the tone for the rest of the year. Academic, behavioral, and social expectations are established, all of which affect the classroom climate. The connections that you make with your students contribute to building a safe and supportive environment in which they feel comfortable taking risks. Such an environment facilitates students' learning of mathematics. Hence careful *planning for the first week of school is essential.* (Chapter 3, "Creating a Productive Classroom Environment," provides further information.)

35 What do students need to know at the beginning of the school year?

Teachers' specific expectations and ways of organizing the classroom differ, but there are several things that are important to communicate at the beginning of the year. Students need to learn what topics they'll study in their math class. They also need to learn the rules or guidelines that support their success in learning math, as well as the specific logistics and routines that keep the classroom running smoothly.

Topics of Study

By the time they reach middle school, most students know that they'll learn more than just arithmetic, that they'll also learn about geometry, probability and statistics, algebra, and so on, but making posters or a chart to hang in the classroom referencing the specific topics they'll study is still useful. A chart might list the following, for example:

Number and Operations

Patterns, Functions, and Algebra

Geometry

Measurement

Statistics and Probability

Each day, refer to the list when introducing lessons. Tell students the mathematical focus both as you begin each lesson and again at the end of the discussion about the activity. A teacher told me that she learned to do this after being observed by her principal during an evaluation. Her principal noted that students could easily explain what they were doing—in this case they were making a coordinate graph—but none of the students knew *why* they were making graphs, how graphs are used, or in what situations a graph would be a useful tool. After reviewing the lesson with her principal, the teacher decided to craft a student-friendly question that encompasses the main mathematical idea of each lesson or activity to help communicate it to students. Say you are teaching a lesson in which students are analyzing patterns in graphs. You might ask, "How can you use a graph to make predictions?" Or if students are engaged in learning about probability, you might pose the question, "What is the difference between experimental probability and theoretical probability?" At the end of the lesson, students discuss the answers and write about their understanding of the mathematical ideas contained in the lesson, which helps inform your subsequent instruction. *(See Chapter 6 for more on writing in math class.)*

Classroom Rules and Guidelines

Guidelines tell students how they are expected to engage in and communicate about their math learning. Some students are surprised to learn, for example, that you expect them to be actively engaged in the task, to think and reason (for some students, adolescence seems to be a stage in which thinking is to be avoided at all costs!), and to share your thinking, both during discussions and in writing. Some teachers create a short list of guidelines and introduce one to students each day. Luis had a list of guidelines on his wall that included the following:

- Respect each other's ideas, questions, and thinking.
- Ask your classmates for help when you need it; help others, without doing their work for them.
- Accept responsibility for your work and behavior.
- Strive for equal participation in groups.
- Everybody cleans up!

During the first week of school, he introduced one guideline each day in the context of the math problem they were doing and had the students discuss what it meant and how it translated into actual practice in the classroom. He had learned from previous experience that his students had difficulty learning all of the guidelines in one day. By contrast, introducing them gradually and reinforcing them daily for the first several weeks of school paid big dividends.

Other teachers prefer to have students generate the rules for the class during the first week of school. While this takes more time, these teachers believe that students have a greater stake in following the rules or agreements when they help create them.

You might give a quiz on the rules at the end of the first week or the beginning of the second week to help communicate to students their importance.

Logistics and Routines

Students need to know how the class is organized for learning. Many teachers find it effective to introduce guidelines or routines in the context of the math problem or activity being discussed. For example, certain lessons will lend themselves to modeling how manipulatives and calculators are to be accessed and when to use them (see Chapters 7 and 8 for further information), the proper procedures for completing a daily warm-up (pages 26–28 give more information about warm-ups), or what the expectations are when working in a small group, with a partner, or by themselves (Chapter 3 provides more information about working in groups). Each day's math activity or problem then provides students with the appropriate context for learning about expectations and routines.

What are some of the things that I should consider when planning my goals for the first week of school? What general preparation is useful?

36

Each teacher wants to accomplish different things during the first week, and it can be helpful to think about the goals you have for students and what you personally want to accomplish. Some initial goals for students might include:

1. Get to know one another and begin to work together as a class community.
2. Learn about expectations with respect to routines, classroom rules, social behavior, and academics.
3. Set up and organize math notebooks.
4. Do math!

As a teacher, you might consider the following goals:

1. Learn everyone's names and something about each student by the end of the week.
2. Learn something about what each student knows mathematically via informal assessment strategies.
3. Connect with parents in some manner.
4. Have students do meaningful math each day.

The goals you set will then dictate what needs to happen before the first day of school. Regardless of the specific goals you set, there are a variety of things that need to be accomplished before students walk in the door for the first time. For many teachers, these include setting up the physical classroom and dealing with logistics.

The Physical Classroom

Setting up the classroom involves the following:

- Determining the desk/table configuration and making a seating chart and nametags
- Organizing student instructional materials (such as texts, math journals, etc.) for distribution
- Putting manipulative materials and calculators in accessible locations
- Making bulletin boards that have designated spaces for student work
- Figuring out a paper-flow system to keep track of papers students turn in and get back from you

Logistics

Logistics involves such things as:

- Writing yearlong plans and daily plans
- Setting up a grade book, attendance book, or other class lists
- Writing and duplicating (1) a welcome letter for students and families, (2) your grading policy, (3) classroom expectations, and (4) a list of supplies each student is expected to bring to class
- Preparing information for substitute/guest teachers
- Becoming familiar with the school handbook (or other document that explains procedures such as the bell schedule, how to assign lockers, the school calendar for the year, the duty schedule, fire-drill procedure, and so forth)
- Finding out who your special education students, English-language learners, and gifted students are and the supports available to these students

37 What might the first week of school look like?

There isn't one right way to open the school year. It's helpful to hear how experienced teachers have launched their first week with students. My colleague La Dona shared with me her general plan for the first week of teaching middle school math (note that in her district, the first day of school is always on a Tuesday, so the outline below covers four days only).

Day One

Materials

nametags (made by students on the first day of school)

classroom rules (these might be duplicated for each student to keep in a notebook as well as written on a poster)

handouts (mixer activity, homework, materials list)

seating chart

To Do:

- *Have students fill out nametags.* La Dona cuts nametags out of tagboard and folds them to look like little tents. She puts the blank nametags out near the door so each student can pick one up on the way in the door on the first day of school, and sit where they like the first day. She has students write their names on their nametags. Having nametags on students' desks helps with taking roll and learning students' names.

- *Fill in seating chart.*
- *Communicate expectations for recording homework assignments.* La Dona models her expectations by recording the first day's assignment on the board; she has students record the assignment in their notebooks.
- *Discuss one classroom rule and have students record it in their notebooks (if duplicated rules were not distributed to students).*
- *Do a mixer activity.* La Dona has her students play *People Bingo (see page 58)* to help students get to know one another.
- *Begin learning students' names.*

Homework

Getting to Know You (see page 59)

About My Student (for parents) *(see page 60)*

Memorize the first classroom rule

Bring classroom supplies in by Friday

Day Two

Materials

handouts (*Grading Policy and Classroom Expectations*, math activity)

To Do:

- *Have students pick up nametags as they enter and either let students choose their own seat or assign one* (see Questions 31 and 32).
- *Make sure students are aware of and follow the first classroom rule.*
- *Collect homework.* La Dona gives students credit not only for their homework but for the *About My Student* assignment their parents were asked to complete.
- *Review how to record homework assignment.*
- *Discuss grading and expectations.* The *Grading Policy and Classroom Expectations* handout (which students are asked to take home to their parents) explains the criteria La Dona uses in grading students' work, and provides a general outline of her expectations for the year *(see pages 172–74)*. La Dona explains the document in class and answers any questions that students have.
- *Introduce and discuss the second classroom rule.*
- *Continue getting to know students' names.*
- *Do math activity* Color Tile Logic *(see pages 61–62)*.
- *Assign homework.*

La Dona knows that getting students involved in mathematical thinking is important during the first week of school; an activity such as *Color Tile Logic (see pages 61–62)* can head off comments such as "I'm not going to learn anything new this year" or "I already know this stuff." She wants to stimulate students' curiosity and engagement. Furthermore, observing students at work on such a task helps her learn what her students know and can do. She uses the math activity as a vehicle for introducing expectations about group work and the use of classroom materials such as color tiles. To reinforce the class rules, after students have played a few rounds of *Color Tile Logic,* La Dona initiates a class discussion. She asks students:

1. Did you remember to explain to the students in your group your reasons for asking the Tile Arranger about certain rows or columns? Did your group get better at explaining their reasoning as the game went on? What could you have said to encourage others to explain their reasoning?
2. Why is it sometimes difficult to understand the reasoning of another student? What is one thing that you could say or do to help you better understand someone's explanation?
3. How did you help others without doing their work for them?
4. What mathematics did your group need to use to do this activity? What did you learn?

By talking about these questions, La Dona communicates her beliefs and values to the class. She sends the clear message that working together to learn mathematics is important and that it matters how students interact with one another.

Homework

Grading Policy and Classroom Expectations handout (Take home for parents to sign; return signed form tomorrow.)

Memorize the second classroom rule.

Bring classroom supplies in by Friday.

Day Three

Materials

Name Values graph

overhead of dollar values for alphabet

To Do:

- *Review classroom rules introduced so far and how to record homework.*
- *Introduce and discuss the third classroom rule.*
- *Do math activity* Name Values.

La Dona likes to engage students in doing the activity she calls *Name Values*, because it gives students practice with computation, provides a context for exploring measures of central tendency, and can be generalized. In the course of doing this activity La Dona also learns about her students' number sense, algebraic reasoning, and problem-solving skills. Furthermore, the activity provides a natural context in which students can learn one another's names. Here's a plan for *Name Values*:

1. Explain the activity: *If A is worth one dollar, B is worth two dollars, C is worth three dollars, and so forth, mentally figure the value of your first name.* (It helps to have an overhead transparency with the values of each letter of the alphabet.)

2. After students have had a couple of minutes to mentally compute the value of their first name, have them share with a partner the value of their name and how they computed mentally.

3. Ask for volunteers to explain how they figured their names' value. This sends the message that you care about their solution strategies and that communicating their thinking is important. After each person shares, ask who else used the same approach.

4. Ask who thinks that they have the "least expensive" name in class, and whose name is "most expensive." Have students mentally find the difference between the two values and ask if anyone knows the vocabulary word for the difference between the lowest and highest value in a data set. (They should be able to tell you that the word is "range.")

5. Tell the class that they're going to make a chart of their names and do a statistical analysis of the data.

6. Using a bold marker, have each person write his or her first name and its value on a large sticky note. Ahead of time, prepare a chart on the whiteboard or a bulletin board for students to post their sticky notes. The dollar values ($1–5, 6–10, 11–15, and so on to $150) are posted next to each other across the middle of the bulletin board so there is space both above (and below) the values for students to post sticky notes with their names and a dollar amount written on it.

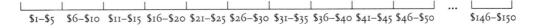

7. Have students come to the board and place their sticky notes in the correct place above the chart.

8. After the chart is complete, ask students what observations they can make about the data on the board. Expect a variety of answers. Remind students what the range of the data is. Ask them what the mode, or most frequent value, of the chart is, eliciting from students what "mode" means. Ask them if it would be possible to find the mean value of the names in the class. Have students calculate the mean.

9. Continue the discussion by asking what the "median" of a set of data is. Ask how to find the median and have several students come to the board and rearrange the names if needed in order to find the median.

10. Next, tell the class that you are going to change the dollar values for the letters. If A is now worth twenty-six dollars, B is worth twenty-five dollars, C is worth twenty-four dollars, and so on, what would the chart look like if everyone recalculated the value of their first names using this new system? Give students a few moments to think privately, and then have them take turns to share their ideas and reasoning with everyone at their table. Then ask for volunteers to share their thinking with the whole class. Typically, students say that the chart will look the same, only reversed.

11. Have students recalculate the value of their first names mentally using the second system, and post their new name values *below* the dollar-value line on the chart, so you don't need to make a new value line or chart.

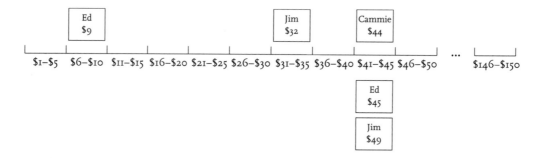

12. Hold a class discussion about what they observe about the data and their inferences about why it looks the way it does. Students are often surprised by the shape of the data.

13. Finally, for older students, tell them that there is a way, if you know the value of someone's name using the first system, to figure out the value of their name using the second system without going through the process of finding the value of each letter and adding. Challenge the class to investigate this problem. It encourages them to think about the idea of generalization. This is a good optional homework assignment. Be sure to check in with students the next day to find out who tackled the problem and how they are thinking about it.

Homework

Memorize the third classroom rule.

Name Values activity extension (This would be a brief inquiry related to the *Name Values* activity. For example, you might ask students to find ten words that are worth from $1 to $10. Can they use all ten in a coherent sentence? Or ask them to predict if there are any values between $1 and $100 that they think will be impossible to find words for. Encourage English-language learners to find words in their native language. Another idea is to ask students to find a sentence that is worth $500, $1,000, $1,500, or some other amount.)

Bring materials in by tomorrow.

Day Four

Materials

math textbooks

handout on notebook organization

To Do:

• *Review all rules learned up to this point.*

• *Introduce and discuss the fourth classroom rule.*

• *Pass out math textbooks and record numbers, have students write their name in ink in their book.*

• *Go over handout on how to set up their notebook, step by step.*

Homework

Study for classroom rules quiz on Monday.

La Dona's colleague Kalusha follows a similar plan, but instead of introducing classroom rules that she's created, Kalusha has her students generate a list of "agreements" that they would be willing to live by over the course of the year. She gives them the following prompt: *To create a safe and productive learning environment in which to learn mathematics, we agree to: _____.*

Kalusha also has students write a math autobiography, or "mathography," in which they describe their past experiences with math. Before doing any writing, she invites students, working in small groups, to describe themselves as a student of math, to explain how they feel about the subject and why. Students are encouraged to talk about which math topics come easily to them and which they find difficult, and to describe what learning math was like for them in previous years. Kalusha has found that students are often more willing to share in small groups than in front of the entire class, especially at the beginning of the year, and being in a small group encourages everyone to speak. She uses a *Go Around* protocol to promote equitable participation in the small groups. *(See Question 24 for more information about protocols.)*

Go Around *Protocol*

1. Decide who will be Student A. That person shares their thinking, ideas, and/or questions.
2. While Student A shares, everyone looks at Student A and listens without interrupting or giving hints that they agree or disagree.
3. When Student A is finished reporting, Student B shares their thinking, ideas, and/or questions.
4. While Student B shares, everyone looks at Student B and listens without interrupting or giving hints that they agree or disagree.
5. Repeat until everyone has shared.
6. Discuss what was shared.

In the past Kalusha has coordinated this activity with the language arts teachers in her school, who help students with the writing process. Finally, students share their mathographies with their table groups. The mathography is a powerful way for Kalusha to learn about her students' relationships with mathematics and their feelings about the subject. This information helps Kalusha be a more effective teacher as she finds ways to connect with students, especially those who express a dislike for math. A word of caution, however: Check with your students' other teachers to find out if they are assigning something similar in their classrooms. Adolescents quickly lose interest in assignments that they find repetitive. An interesting alternative assignment would be to have students respond to the question, What *does* it mean to be good at math? Answers to this question will provide you with rich insights into their experiences with math.

38 It's only the first week of school and already I feel as if I'm losing the interest of some of my students. Some even say that they hate math! How can I turn things around?

It's unfortunate, but by the time some students reach adolescence, they have been turned off of math. Students communicate to us in a variety of ways that they dislike math. Sometimes they are direct, which is helpful, because by acknowledging their feelings they give us an "opening" to find out why they feel this way. Having students write a math autobiography at the beginning of the year allows your students to describe the kinds of experiences they have had with math, which can give you valuable information about their needs *(see Question 37)*.

More often than not, if students feel negatively about math, it is communicated indirectly, through their misbehavior. Misbehaving in class can be a clue to teachers that a student fears or dislikes math. Some students file into my classroom at the beginning of the year and immediately start to act out. Sometimes I wonder if this is done to deflect attention away from what the student doesn't know mathematically and fears I'll find out. When a student offers elaborate and varied excuses for not participating in class or not doing their assignments, I'm alerted to the possibility that the student is trying to conceal serious math deficits. In extreme cases, I've had middle schoolers confess to me that they'd rather misbehave and be sent to the principal's office than have to sit through math class!

You need to find out why some students dislike math and to acknowledge their feelings. For too many adolescents, their math learning has been a steady diet of abstractions and disconnected skills, damping students' enthusiasm and curiosity. Imagine what that must feel like for a student. To draw an analogy to sports, it's like being told over and over how to hit a baseball without being allowed to pick up a bat. Or, with respect to the arts, imagine being taught endlessly how to mix colors, but never being allowed to use them to paint. Or being required to practice scales over and over but never having the opportunity to play music. When students are taught only abstractions and skills, it's not surprising that they grow to dislike math. To counteract these kinds of experiences, it's critical to provide activities that make the math accessible to students, that engage their curiosity and enthusiasm, that encourage them to make sense of math, and that allow them to experience the breadth of the subject so that they learn that math has real-world relevance. The *Color Tile Logic* activity described in Question 37 is an example of such an activity, and others are sprinkled throughout this book. Not every activity will excite every student, of course, but what students do in your math class should engage them as much as possible in thinking, reasoning, and solving problems.

Sometimes it's peer pressure that causes adolescents to say that they hate math. "Nobody cool likes math!" some claim. I smile when I hear comments like this and have lots of comebacks in my "math is cool" arsenal that resonate with young people. For example, I give examples of famous people who studied math in college, such as Michael Jordan. I thank students for being honest with me about their experiences, and for letting me challenge them to see if I can change their beliefs. I tell them that I'll check in with them at the end of the year and see if they still feel the same way. Thankfully, in my experience, most of them don't.

Sometimes students' negative feelings about math are "handed down" to them from their parents, who had bad experiences with learning math in school. Most worrisome to me is when students hate math because they've gotten messages over time that they're not good at it. These are the students whose past experience with math has been a steady regimen of drill and practice worksheets that has convinced them that they are not competent or capable of doing mathematics. In these situations, students' confidence is understandably low. People customarily feel uncomfortable in situations in which they lack understanding, especially when they are in the minority. Think about a time when you were among a group of people who were discussing something that you knew little about, and you felt that you *should* have known more about the topic. Not a good feeling. Middle schoolers who haven't experienced success in math class can feel powerless, stupid, or worse. Often they feel that they have missed the boat with respect to math and will never catch up with their classmates. In these cases, it's important to let students know that you are aware of how much they're struggling, that you're committed to doing your best to help them learn, and that you're confident that they *will* learn—and even come to love— mathematics. For this to happen, they need learning experiences that are accessible, that get them actively engaged in the subject, and that allow them to work from their strengths and interests, since the key to changing students' attitudes is success.

Sometimes students will tell you that they are simply not very interested in math. When I hear comments like this, I ask students what subjects they *are* interested in (this is where a math autobiography, or "mathography" *[see Question 37]* can be helpful). I think about ways to connect the math I'm teaching to those interests. If, for example, a student tells me that she loves art, I look for geometry activities that incorporate art to support her math learning. Similarly, if a student expresses that he likes writing, I remind him of the many opportunities for him to write in math class. *(See Chapter 6, "Incorporating Writing into Math Class," for further information.)* Let such students know that you'll work with them to help them discover that mathematics is both interesting and enjoyable.

39 I'd like to connect with my students' parents during the first week of school, but I don't have time to call all of them. What are some ways I can reasonably communicate with students' families?

It's a good idea to connect with parents as early as possible in the school year. I think about communicating with them in a variety of ways: in person, on the phone, by e-mail, via a Web site, and through written letters and notices that are sent home. Generally speaking, the first week of school I send home information about our class, the homework expectations, and my grading system. *(See Chapter 10, Question 122, for a sample letter.)* I post the information on our school's Web site as well. I also send home and post a list of supplies that students need to bring to school.

I call parents, beginning on the first day of school. To limit the time encroachment into my personal life, I make only two or three calls a night. I begin by calling the parents of students who stand out in some way. I may be concerned about a student's behavior or their level of proficiency (if they seem to lag behind or are far ahead of their classmates). I introduce myself to the parent and let them know I'm contacting the families of each student in my class. I always say something positive about their child, and then I ask them to tell me a bit about their adolescent and what I should know about the student. Parents usually appreciate the inquiry and are more than happy to fill me in. I use a journal and take detailed notes for future reference about the call, including the date and with whom I spoke. I don't make complaints about the student during the first phone call but instead work on building a relationship with the parent, knowing that if we have a good relationship, tackling problems together will be much easier. I also invite parents to Back-to-School Night and ask them to put the date on their calendar. Finally, I thank the parent for their time and let them know that I'm looking forward to a great year with their child.

These calls are important. Not only do I learn valuable information about my students, but in the event that I need to contact the parents about academic or behavioral issues, the fact that I've already begun to establish a relationship with them makes the conversation more productive, since the parent knows that I have their child's best interests at heart.

Additionally, I often send home a letter telling parents about our class and what students will be learning for the first few weeks of the year. *(See Question 122 in Chapter 10 for a sample letter.)*

During the first week of school, I also send home the following questions for parents to respond to in writing or via e-mail that gives me information about their child. I ask parents:

1. What are your child's strengths as a learner? His or her strengths in math?
2. What is challenging for your child as a learner? What challenges him or her in math?
3. What else should I know about your child?

Parents tell me that they appreciate being asked, and I get a wealth of information about my students that helps me be a better teacher.

Calling parents, sending letters home, and posting information on Web sites requires time and energy, which are in short supply for many teachers. But the benefits are well worth the effort for everyone—students, students' families, and teachers.

PEOPLE BINGO

Rules

1. Find a person for whom a statement is true.
2. Have him or her sign name in the box. Learn one another's names.
3. You can use the same person's name two times.
4. Try to fill in the whole sheet!

1. has at least one pet	2. has more than eight letters in first name	3. loves math	4. has the same birth month as you	5. can tell you the months with an *r* in them
6. can count to 5 in more than 1 language	7. likes hip-hop music	8. can name 5 geometric shapes	9. is taller than 150 cm	10. blue is his or her favorite color
11. has more than 6 people in his or her family	12. knows something about Fibonacci	FREE: PUT YOUR NAME HERE	14. plays on a sports team	15. can solve: $\dfrac{2}{3} \times \dfrac{1}{4}$
16. favorite subject is the same as yours	17. knows the value of pi to the hundredths	18. can list 5 prime numbers	19. went to a different state in the past year	20. is an only child
21. has used pentominoes	22. has his or her own calculator	23. likes to play board games	24. thinks he or she can beat the teacher in a math game	25. can do 30 x 25 in his or her head

GETTING TO KNOW YOU

Three things I like
to do with my
family are:

1.

2.

3.

Here's a picture
of me:

My favorite books
are:

1.

2.

3.

My favorite movies
of all time are:

1.

2.

3.

4.

In my free time, I like to:

If you could have
dinner with anybody
who ever lived, who
would you choose?

_____

I think I'm old enough,
but my parents think
I'm too young to:

1.

2.

3.

My favorite songs are:

1.

2.

3.

Dear Parent,

I'd like to get to know your student as quickly as possible this year. Would you please write answers to the questions below and send them back to school with your child tomorrow?

Thank you.
Sincerely,
Tilly Teacher

About My Student

My child's name (please print) _____

My name _____

1. What is important for me to know about your child?

2. What does your child like to do in his or her free time?

3. How does your child feel about math?

Color Tile Logic

You need:

a group of three or four
color tiles
grid paper

To do:
Read all of the directions aloud before beginning.

1. Decide who will be the Tile Arranger. This person takes nine color tiles (three of each color) and arranges them in a secret 3 × 3 (or 4 × 4) grid. Tiles of the same color must be placed so that they are touching, either side to side or corner to corner. A sample secret grid may look like this:

	Column A	Column B	Column C
Row 1	Red	Red	Red
Row 2	Blue	Blue	Green
Row 3	Blue	Green	Green

The object is for the group to use logical reasoning to figure out how the color tiles have been arranged by the Tile Arranger in the 3 × 3 grid. The group should determine the answer by asking as few questions as possible.

2. The group talks about which column or row they want to know about, and each person explains why he or she wants to know about that row or column. The group should decide together which one and say, "Tile Arranger, tell us about

_____   _____."

(column or row)   (letter or number)

Continued on page 62

Continued from page 61

3. The Tile Arranger tells the group two things:

    a. what colors are in the selected row **or** column

    b. how many tiles of each color are in the row or column (but not the order in which the colors appear!)

4. The group thinks about the answer and then decides:

    a. what is known about the grid for sure and why

    b. what the possibilities might be

5. Each person in the group takes turns and talks about which column or row they want to know about next, explaining why they want to know about that row or column. The group decides together which one and says, "Tile Arranger, tell us about

    _____   _____."

    (column or row)   (letter or number)

6. The Tile Arranger tells the group two things:

    a. what colors are in the selected row **or** column

    b. how many tiles of each color are in the row or column (but not the order in which the colors appear!)

7. Repeat. Each time the group guesses, mark a tally to keep track of the number of guesses it takes to solve the problem.

8. When the group has enough information, they name the pattern and the Tile Arranger shows the secret grid.

9. Rotate so the person who is sitting to the left of the Tile Arranger becomes the new Tile Arranger for the group and play again.

While you work, your group is practicing:

• discussing and deciding together

• giving reasons for your suggestions

• helping others without doing their work for them.

When time is called, be ready to talk about how your group did.

Five | *Leading Class Discussions*

LaToya, Barbara, and Karl were talking together after the district in-service session.

"The facilitator at today's meeting was clear that discussions in math class are important," LaToya observed, "and I want to try out some of the techniques she used to get kids to talk with one another after they've worked on a problem and before they write about it. My students are willing to work hard in class, but lately I've been avoiding asking them to talk about what they are learning because it seems like it's always the same five students who volunteer and the discussions don't go anywhere. It's frustrating."

"I hear you," replied Karl. "The same thing happens with my students. But I'm more concerned that in my class, when students begin sharing what they learned, someone always presents a really sophisticated approach and the other students just seem to shut down. They don't want to contribute to the conversation after that."

"I don't really have problems with class discussions," Barbara offered. Both LaToya and Karl looked at her in surprise. "Because there's never any time in my class to hold them!" she explained. "My students seem to need all the time just to finish their work."

Leading class discussions is one of a teacher's most important jobs. It requires careful thinking before the lesson is taught because the kinds of discourse that a teacher expects from his or her students influences what they learn. Many teachers today expect their students to explain what they did and how they solved various problems. In order to maximize learning, however, it's important to press students to justify and generalize their ideas. When you devote specific attention to discussions, students learn that they have the responsibility to pay attention to their classmates' ideas, and make sense of, evaluate, and interpret those ideas.

40 What kinds of questions lead to quality class discussions?

For class discussions that focus on students' thinking and press for understanding, use the following general questions to solicit their responses:

- *How did you figure out your answer?* It's important for students to explain their reasoning *every time* they give an answer to a problem. Students' explanations are critical windows into their thinking processes and give you valuable insight into their level of understanding. The expectation to explain their reasoning should be internalized by students, so that they habitually include explanations with their responses without being prompted.

- *How do you know your answer is mathematically reasonable?* This question helps children revisit a problem and reconsider their solution. Judging reasonableness, explaining why by providing mathematical reasoning, and verifying solution processes helps students build their math sense. Furthermore, when status issues are operative in the class *(see Chapter 3, Question 26 for more about dealing with status issues)* students need to learn that the "correctness" of a response lies not in the social or academic status of the student who offered it but in the quality of the mathematical argument (Hiebert et al. 1997).

- *Who can explain what _____ said in your own words?* Asking students to explain a classmate's idea is useful. It encourages students to think carefully about and process what was said. Having various students state an idea in their own words provides opportunities for students who might not understand one explanation to find meaning in another. It reminds children that there are different ways to express the same idea. Furthermore, having students paraphrase what other students said reinforces the expectation that they listen carefully.

- *What questions do you have about _____'s strategy? Talk with a neighbor and then be ready to share.* It's important to provide students opportunities to challenge or defend the mathematical validity of ideas. You want to encourage them to employ their mathematical reasoning when thinking about strategies. Having students talk to a neighbor about questions they have gives them the opportunity to check out their thinking in a safe environment before sharing with the class.

- *Why does _____'s idea make sense mathematically?* Pressing students to explain their reasoning and justify their ideas supports their mathematical understanding. Students learn to determine the "correctness" of a solution or method based on its mathematical merit rather than on the social or academic status of the individual contributor.

- *How are ____'s and ____'s ideas alike mathematically? How are they different?* Asking students to compare and contrast mathematical ideas encourages them to make connections between ideas and to justify their responses.
- *What new ideas does this problem cause you to think about? Would this happen in every case? How might you prove it? Take a few minutes to think about this before raising your hand to share.* It's valuable to explore the new ideas that are sparked as students share their thinking during a class discussion. Furthermore, students need to be regularly asked to make conjectures and predictions as well as explain and justify what will happen in the general case or in different contexts. This helps students make connections between big ideas. Students also need many opportunities to engage with the concept of proof in mathematics.
- *Raise your hand if you used the same strategy as ____ did.* This validates those students who approached the problem or activity in the same way as the student who presented the idea. It also gives you a sense of how common the strategy was. Ask this question after each student presents.
- *Who has another way to solve the problem (or approach the activity)?* Students should learn that there are often many different solution strategies. Making sense of different ways to solve problems promotes flexibility in thinking. Occasionally a student might say, "My way was sort of like ____'s way." Encourage her to explain her method in her own words. The more explanations there are, the more likely it is that all students will understand the point under discussion. *(See Suzanne Chapin, Catherine O'Connor, and Nancy Anderson's* Classroom Discussions: Using Math Talk to Help Students Learn *[2003] for further information.)*

41 How can I help my students become better at participating in class discussions?

First, you need to hold such discussions regularly. Then, the best way to help your students to successfully participate in discussions is to share your expectations and then model and reinforce them regularly in the context of class discussions. Linda, for example, made a poster of the following guidelines for class discussions and shared them with her class:

- *Share your ideas in turn and explain your reasoning. It helps all of us learn.*
- *While someone is presenting to the class, stay seated, look at the presenter, listen to them, and think about what they are saying. If you notice that you are not paying attention or your attention is wandering, write some notes about what the speaker is saying to refocus your attention.*

- *Listen carefully to what someone is explaining. Do their ideas make sense? Why or why not? Raise your hand and wait for the presenter to call on you. Then explain what their thinking sparked for you, or ask them a question.*
- *At the end of a discussion, be ready to summarize what you have heard and/or learned, ask a question, or make a conjecture about the mathematical idea.*

Linda learned that her students got better at participating in discussions when she asked open-ended questions. Linda used to ask questions such as *What's the answer?* Now she says to her students, "On the count of three, let's all whisper the answer." As soon as students do so she says, "Okay, now let's get to the really interesting part—how you figured it out!"

42 What's a good way to introduce discussions in my class?

Model your expectations for discussion by offering students firsthand experience. For example, at the beginning of the year a teacher explained to her seventh-grade class that in math class that year, students would have many opportunities to share their answers and thinking with the class to help them learn. She then put three scoops of rice into a jar, so that it was about one-third full, and asked the students to estimate the number of scoops it would take to fill the jar to the top. Many hands shot up. The teacher was interested in hearing how students would use the information to make their estimates. But when called on, students typically just reported a number. After each estimate, however, she asked, "Why do you think that's a reasonable estimate?" After posing this query to several students, she shortened the prompt to a simple, "Because?" Finally, students caught on and began to offer explanations without additional encouragement. After the last student had an opportunity to share, the teacher said, "In this class this year, I'm interested in your answer and especially in your thinking. When I ask you questions in class, and you give an answer, be ready to explain why you think that. It helps you clarify what you mean, and other students hear strategies and approaches that they might not have thought about, or you may say something that confirms their thinking, and I learn how you think." The teacher then filled the jar to the top. It took eleven scoops of rice.

The teacher then told the students that she was going to empty the jar, use the same scoop, and fill it a second time, but this time with beans. She asked the class, "Do you think there will be more scoops, fewer scoops, or the same number of scoops of rice as beans? Talk with a neighbor." There was a flurry of conversation, and after a few moments the teacher called the class back to attention. Several students blurted out their estimate. The teacher quieted the class and told the students that she wanted them to talk one at a time so that they could hear one another's ideas. She reminded them to raise a hand to let her know that they wanted to contribute. Before she called on a student to respond, she gave another reminder: "When someone is talking, listen carefully to see if you agree or if your idea is different."

She called on Karen first, asked the rest of the class to lower their hands, and waited until they did so before signaling Karen to speak. Karen then said, "I think there will be fewer beans." Other hands then shot up. The teacher asked them to lower their hands again and, turning to Karen, she prompted, "Because?" Karen looked a bit surprised by the teacher's prompt; she still wasn't used to explaining her answers. She thought for a moment and said, "Because the beans are bigger than the grains of rice, it will take fewer scoops to fill the jar." "How many of you agree with Karen that there will be fewer scoops of beans than there were of rice?" asked the teacher. Hands shot up again. She waited a moment and asked, "Who has a reason that's different from the reason Karen gave? Karen, would you please call on someone?" In this way, she reinforced that students were to pay attention when their classmates offered ideas. Often in classroom discussions, students merely wait for a classmate to stop talking so that they can have their own turn. However, communication works best when students actively listen and respond to one another.

The discussion continued until all students who wanted to share had a chance to communicate their thinking. The students' different points of view contributed to a lively discussion. The teacher then had the students write about how many scoops of beans they thought would fill the jar and why.

This particular lesson was done with a group of seventh graders, but it works well in other grades as well. It's a good example for showing how to make discussion guidelines explicit in the context of an actual lesson. It's also a good lesson for this purpose because it engages students' curiosity and doesn't require complex thinking that would exclude anyone. Finally, don't forget to follow through with the inquiry once students have made their estimates, to prove how many scoops of rice or beans the jar actually holds!

After the class has had several experiences with whole-class discussions, you might post your guidelines for discussion as a reference.

43 **I'm comfortable giving my students problems to work on in pairs or groups and circulating around the class to help and guide them. But I'm not comfortable leading discussions with the whole class. I always seem to run out of time. Are whole-class discussions really that important?**

Class discussions are essential. They provide time for students to share what they learned, ask questions, consider new ideas and other perspectives, and get feedback about their thinking from classmates and from the teacher. They are extremely helpful before students are asked to write about their learning. Discussions can take place when you're introducing a new topic or unit, to provide clarification in the middle of an exploration, and during the summary at the end of an investigation.

Class discussions provide opportunities for teachers to assess instruction as they support students' thinking. They can give us a general sense of the class's response to a particular lesson—how it was received and what students understand. Discussions don't replace the need for assessment *(see Chapter 9 for more about assessment)*, of course, but they provide useful information.

Summarizing problems or investigations is important for several reasons. It allows answers and generalizations to become explicit. As students share their thinking, other students benefit from hearing about the strategies used. And class discussions allow you to address misconceptions, make important ideas clear, and raise additional related questions and ideas. Finally, students benefit from many opportunities to talk about their learning before writing about it. *(See Chapter 6, "Incorporating Writing into Math Class.")*

44 I don't want to interrupt my students when they're working on a problem or an exploration, but I also don't want to skimp on class discussions. How do I carve out sufficient time for discussions?

Don't skimp on discussion time. It's best to hold the discussion when the investigation is still fresh in the students' minds, preferably right after they investigate a problem. Cut the exploration short if you need to. If you wait until every student has completed the problem, you'll likely run out of time for a summary discussion. Before students begin work, tell them how much time they have to work on the activity (work the problem yourself ahead of time to estimate the time needed). Set a kitchen timer to chime five minutes before you want the discussion to begin.

If you do find yourself running out of time during any particular class, consider discussing the problem the next day. While it's not an optimal strategy, it's better than moving on to the next exploration without giving your students sufficient opportunity to articulate their thinking. If you do push the discussion to a later day, you'll need to spend time reviewing the exploration with students before you begin the discussion.

45 What happens if I have difficulty following what students say when they present?

Before you initiate a class discussion, think through the problem or activity you've assigned and come up with as many ways as you think students will use to approach the problem. Anticipating student responses is easier to do for lessons you've taught before and more difficult for lessons that are new to you. In both instances, advance preparation is key. At the same time, be open to the fact that a student might use a strategy that you did not expect. While students are working, circulate throughout the room to observe how they are engaging in the activity or problem. Talk to students whose approaches you don't recognize. This will make it more likely that you'll understand a student's thinking during class discussion.

If you have trouble following a student's reasoning, after he has finished sharing ask the class to paraphrase the student's ideas and explain in their own words what the student said. Another method is to tell the student whose reasoning you're having trouble understanding that you need some help following their line of thinking. Can they explain their idea in a different way? Or ask the class if they can explain the student's idea. Doing this not only can help you figure out what's on a child's mind but also models that you value persistence in understanding others' ideas. Be sure to do this calmly and respectfully—it's important that your tone of voice not communicate that you think a student's line of reasoning is faulty; no student should be made to feel embarrassed that they've said something that the teacher doesn't understand.

Sometimes I make the mistake of listening for the answer I expect or want to hear. When this happens, I often misunderstand what the student is saying or I misinterpret their response. Instead I should focus my attention on listening carefully to what students say to understand their thinking. For example, when asking students to explain how to determine the theoretical probability of rolling a sum of 7 on a pair of dice, I may anticipate a specific method. Perhaps I expect an explanation that involves listing all of the combinations of 7 on two different dice:

Red Die	Green Die	Sum	Red Die	Green Die	Sum	Red Die	Green Die	Sum
1	1	2	3	1	4	5	1	6
1	2	3	3	2	5	5	2	⑦
1	3	4	3	3	6	5	3	8
1	4	5	3	4	⑦	5	4	9
1	5	6	3	5	8	5	5	10
1	6	⑦	3	6	9	5	6	11
2	1	3	4	1	5	6	1	⑦
2	2	4	4	2	6	6	2	8
2	3	5	4	3	⑦	6	3	9
2	4	6	4	4	8	6	4	10
2	5	⑦	4	5	9	6	5	11
2	6	8	4	6	10	6	6	12

Thus the theoretical probability is $\frac{6}{36}$, or $\frac{1}{6}$. However, a student may offer a completely different method. There are several ways to explain how to find the theoretical probability of a sum of 7 on two dice. One might solve the problem using a tree diagram:

Red die	Green die	Sum		Red die	Green die	Sum
	1 =	2			1 =	5
	2 =	3			2 =	6
	3 =	4			3 =	⑦
1	4 =	5		4	4 =	8
	5 =	6			5 =	9
	6 =	⑦			6 =	10
	1 =	3			1 =	6
	2 =	4			2 =	⑦
	3 =	5			3 =	8
2	4 =	6		5	4 =	9
	5 =	⑦			5 =	10
	6 =	8			6 =	11
	1 =	4			1 =	⑦
	2 =	5			2 =	8
	3 =	6			3 =	9
3	4 =	⑦		6	4 =	10
	5 =	8			5 =	11
	6 =	9			6 =	12

Or a student might create a chart to show the probability:

number on first die

+	1	2	3	4	5	6
1	2	3	4	5	6	7
2	3	4	5	6	7	8
3	4	5	6	7	8	9
4	5	6	7	8	9	10
5	6	7	8	9	10	11
6	7	8	9	10	11	12

number on second die

Acknowledge the student's thinking and stay open to her thought processes. Remember that valid student responses won't always replicate your own thinking or the examples given in the teacher's guide.

46 How do I decide who to call on during class discussions, so that I don't end up with the same students presenting over and over again?

There are several options for encouraging students to present, and there are pros and cons associated with each. You can call on only those students who volunteer. This allows those students who feel comfortable sharing before the class the opportunity to report their findings, and it can reduce the anxiety level of students who are not yet comfortable sharing in a whole-group setting. But on the other hand, calling only on students who volunteer can encourage a lack of accountability in other students. Students who know you won't call on them might feel that they don't need to attend to the discussion. And this strategy also tends to result in the same students presenting over and over again.

You can call on a specific child by name. While for some students this can be initially uncomfortable until a safe classroom environment is established, it increases the number of students who present and ensures that all students are represented in the discussion. Some teachers write the names of each student on craft sticks or index cards and put the sticks or cards in a can or other sturdy container. After asking a question, the teacher pulls out one stick or card at random and reads the name written on it. That student is then asked to respond to the question.

Another idea is to ask for volunteers, but explain that once someone has contributed an idea to the discussion, they must wait until the class has heard from five other students before raising their hand again. Then ask to hear from someone who has not yet shared his or her thinking.

47 What if nobody volunteers to share during the discussion? How do I get students to speak up?

Students need to understand your expectations for their behavior during a discussion. Tell your students that part of their responsibility as students is to contribute to class discussions. Acknowledge that it's harder for some people to do this than for others, but it's important for everyone to participate. Students know who among them in a class loves to talk in front of the whole group and who finds sharing difficult. Offer students ways to help one another. For example, having students talk in pairs or small groups before whole-class discussions can give reluctant students more confidence about speaking up. Also, be sure to allow enough time for reflection after asking a question. To yourself count slowly to seven, or even ten, before calling on a student.

I've heard of some creative ways that teachers use to encourage class participation, especially when students work together in pairs or groups. My friend John solicits responses by cleverly crafting his prompts. He'll say, for example, "The person with the most pets at home will report" or "Let's hear from the person with the most pockets on their clothes today." John says that such prompts communicate to students that everyone is accountable for sharing their thinking and thus must attend to the discussion. Furthermore, he explains that this approach is less threatening to students than putting them on the spot by calling their names.

If the students are clear about your expectations for class participation and still nobody volunteers, it could indicate that what you asked is unclear. Ask someone to paraphrase your question. Or say to students, "Let me ask my question in a different way," and rephrase the question.

Many teachers take into account participation in class discussions when determining students' grades, to emphasize that class participation is an important and valued part of math class.

48 During discussions with the whole class, sometimes the first student to share offers an idea that's sophisticated or complex, and the other students seem to shut down because their ideas aren't as fancy. Or a child shares something so esoteric, the discussion grinds to a halt. How do I facilitate discussions to keep the momentum going?

It can be helpful to plan your discussion time so that the less complex ideas are presented early on, and the more complex ideas come out toward the end of the summary. When a complicated explanation is presented first, students with less elaborate ways of seeing the problem are often reluctant to share, feeling that their way is inferior. A word of caution, however: Adolescents are quick to catch on when a teacher regularly "rigs" discussions so that ideas are presented in a certain order, so use this approach judiciously. Many times it doesn't matter in what order ideas are presented; what's most important is that the ideas are made public.

Before posing a problem to the class, think about what math concepts or skills are embedded in the problem that are essential to highlight during the discussion. Spend time anticipating how students will solve the problem. What strategies and methods do you think they'll use? This is important because as students work, you can circulate throughout the room and notice the approaches they use. Some teachers keep a clipboard with a class list on it to make notes about individual students' comments, thinking, and strategies. These personal communications are a record you can use when orchestrating the summary of the lesson. Furthermore, by thinking through possible solution strategies, you have a chance to think ahead of time about how you'll respond when students offer various ideas, correct or incorrect, simple or complex. Consider what you will ask or say when students come to consensus but the conclusion is mathematically incorrect or of little mathematical importance. If a student shares something that might be meaningful to him but holds little interest or learning opportunity for the class, acknowledge the student's contribution and then redirect the discussion by asking a question directly related to the problem or inquiry. Knowing what to do with the information students share in a discussion and what questions to ask them to extend, clarify, or rethink their ideas is key to keeping discussions lively and on track.

49 How do I establish a classroom atmosphere in which class discussion is valued? How do I help students learn respectful norms for interaction during discussions?

Be explicit with students in explaining both the purposes of class discussions and your expectations for them during discussions. Do this in the context of whole-class discussions, both at the beginning of the year and periodically throughout the year.

Acknowledge that class discussions might be new or uncomfortable to students who have not experienced them or who have come to expect that the most important thing in math class is to "get the answer." Students need to hear the rationale and purpose for discussions over and over. When students recognize that they and their classmates are learning from the discussions, they begin to value them and participate more fully.

Frequently remind students about the purposes for holding class discussions:

- *Discussions provide you with the opportunity to consider new ideas, which helps you learn.*
- *During discussions, hearing others' ideas can help you think about a problem or idea in a new way.*
- *We need to work together to help everyone in our class learn, and sharing ideas during discussions is one way to do that.*
- *As a teacher, when you share your ideas, it makes me think carefully about the lessons I can prepare to help you learn.*

Remind students that they have several responsibilities during a discussion:

- *When someone is presenting to the class, stay seated, look at the presenter, listen to them, and think about what they are saying—don't talk, work, or wander.* Following through with this expectation is crucial. When they're not interested in a discussion (no matter what the reason), adolescents are masters at getting out of it: they need to use the bathroom, get a drink of water, sharpen their pencil, put materials away, visit their locker, and so on. Telling students up front what your expectations are and then following through consistently is the most effective (some teachers would say the *only*) way to deal with this.

- *As your classmates explain their thinking, your job is to think carefully about their reasoning and conclusions. If you don't understand, figure out what it is you don't understand. Then raise your hand and ask the student who is presenting for clarification.* This lets students know that listening in order to make sense of what is being presented is paramount.

- *Think about what is being presented. How does it relate to or confirm what you know? At the end of a presentation, you want to be able to either paraphrase what the presenter said, add a related idea, or offer a new idea or question. Raise your hand when it seems appropriate and share these with the class. Your ideas will help everyone learn more.* Providing concrete and specific suggestions for how to think about what's being presented is helpful for adolescents. Give your students opportunities to put the discussion guidelines in action. For example, after a student has presented an idea to the class, have the other students talk with a partner about what the speaker said. Or have students write about the presenter's ideas or jot down a question that was sparked as a result of the student's contribution.

Talking in front of the class is sometimes intimidating for adolescents. Thus it's important that they know that it's okay to make a mistake. Point out that mistakes help us learn, and one student's error is an opportunity for everyone to benefit. Building a classroom environment that encourages risk taking and sharing of ideas is essential. The best way to do this is to model for your students how to be respectful and supportive and then insist that they do the same with their classmates.

Some teachers find it useful to record discussion guidelines on a poster to hang in the classroom. Others duplicate the guidelines for students to keep in a binder. My colleague John factors participation in discussions into student's grades. He feels strongly that doing so encourages students to participate.

50 What's my role during a class discussion? What should I tell students about my role?

It's important to model your expectations consistently and clearly. For example, if you tell students that you expect them to present their work to the class, give them plenty of opportunities to do so. When one student is talking, listen carefully to what he or she is saying. If you want the students who are presenting at the board or overhead to address the class instead of focusing on you, consider sitting down in their seat while they talk. Act exactly how you want students to act, asking questions, adding new ideas, or summarizing the discussion, following the protocol you expect your students to follow. This sends a strong message to students. *(See Question 48 on page 72.)*

51 My students ask me why they have to explain their thinking all of the time. How should I answer them?

Answer them clearly, concisely, and consistently. Tell students regularly that communicating their ideas is one of the key elements in learning. Remind students that when they talk and write about their mathematical ideas, they must organize and make clear their thinking and reasoning. This helps them solidify their understanding. Also remind them that listening to others and considering their ideas and strategies helps them learn as well.

The following prompts are useful ways for encouraging students to explain their thinking and reasoning. The more often you use them, the more quickly they'll become part of your teaching repertoire. Furthermore, students will quickly learn that this is what math class is about.

Explain how you figured that out.

What do you think?

Why do you think that?

Say some more about what you are thinking.

Who has another way to explain what _____ said?

Why do you agree with _____'s idea?

Why do you disagree with _____'s method for solving the problem?

What's another way to explain that? (For variations and elaborations on these prompts, see Question 40.)

52 Occasionally when I ask a student to explain his or her thinking, the student shrugs and replies, "I just know." How should I respond?

This response occurs frequently in classrooms where adolescents haven't had much experience with explaining their thinking. Like most things, however, with practice students get better at doing so. There isn't a guaranteed way to help students recall or re-create their train of thought. Consider gently probing their thinking by asking the following:

> *What was the first thing you thought about?*
>
> *Where did you go from there?*
>
> *Could you draw a diagram to show us what you did?*
>
> *How would you tell someone who was absent (or is in a younger grade) what you were thinking?*

If a student still struggles to offer an explanation, have classmates provide support. Ask if someone else in the student's group or at their table can explain the student's thinking, as a way of modeling for the student the kinds of explanations you're looking for. Have the student comment on his classmate's explanation: Did it explain her thinking? And if not, why not?

Be sure you have this conversation in a manner that is nonthreatening and respectful. If students feel as though they are being unfairly put on the spot (keep in mind that asking them to explain their reasoning is not unfair!), they may shut down and turn off to math. More so than younger students, adolescents can be hypersensitive about what they perceive others think about them, and unwillingness to explain their thinking may be symptomatic of this.

Adolescents can find it helpful to talk with a partner or group before sharing their thinking with the whole class. A small group provides a safe environment in which to take risks and gives more students the opportunity to share. Being able to check out their ideas in a small group is reassuring for many adolescents.

53 How do I help students prepare for presenting their ideas in a class discussion?

Tell students who will be asked to share their work and ideas during the class discussion to think ahead of time about their responses. Have students practice first in pairs, if time permits. Encourage students to explain their thinking step by step, rather than assuming that the class knows how they arrived at their ideas. Consider telling students:

- *If asked to present, think about what to share and in what order to share it. You may be given overhead transparencies and pens to record your work for the class to see, or be asked to write it on the board.*

- *If you're presenting and get stuck, ask the class—not the teacher—for help.*

- *Call on fellow students when you see someone with their hand raised—don't wait for the teacher to do it.*
- *When presenting to the class during the discussion, look at and talk to your classmates as well as to the teacher. The idea is to have a conversation with the class as a whole, not a private conversation between you and the teacher held in front of the class.*

54 I find it useful for students to show their work on an overhead when they present, but this process can be time consuming. My class period is only forty-five minutes. How can I maximize time for such presentations?

Have students prepare overheads ahead of time. Not only does this allow for better flow of presentations, preparing transparencies independently gives students a chance to reflect on and edit their work. When using the overhead, select ahead of time which students you want to make presentations, and pass out blank transparencies and pens to these students during work time.

Good alternatives to the overhead are chart paper and newsprint. Post these at the beginning of work time and ask students to record their results and strategies; these can then be referred to during class discussion. Students can also be asked to record their work on the board. Posting allows students to refer back to a particular piece of work as needed during the subsequent presentations. The disadvantage of using an overhead is that once a student has presented their work and removed the transparency, it is no longer available for use as a reference. Overheads are useful, however, if you want ideas introduced into the discussion in a particular order.

Some teachers use document cameras in their classrooms. Students simply place their written work in the appropriate place under the camera, where it is then projected onto a screen. Using a document camera has similar advantages and disadvantages as an overhead projector.

55 How do I help my students stay on task and focused during a class discussion?

During a presentation or after someone has presented, ask students to turn to a partner or to their group to try to restate what the presenter has said, while considering if it makes sense to them. The processing time is necessary! Alternatively, ask students to write about the idea they just heard. Or, you could have students try the strategy on a new problem to see if it works and makes sense before going on to the next presenter. Some teachers have their students take notes during discussions to help them make sense of the ideas shared and for future use.

56 I worry that students who struggle with math will find class discussions confusing and unhelpful. I've seen students who barely grasp one strategy simply tune out when other ideas are presented.

Occasionally students will tune out during a class discussion. This can happen because students are confused by what's being said or because they're not interested. Adolescents have active minds, which can sometimes work against you during class discussions! Also, it's sometimes hard to follow someone else's reasoning. Think about times when you thought you were giving an extremely clear explanation to students about why something makes sense, only to elicit blank stares in return. While children's confusion might cause you to consider reducing the amount of time you spend on class discussions, more time might in fact be the answer.

We know that when we teach something, we have to make sense of it for ourselves—to "own it"—before we can offer it for students' consideration. The same holds true for students. Thus, consider asking students in partners or groups to paraphrase an idea shared during a discussion. You might also ask students to use an idea presented in a new situation. For example, if a student shares how they convert fractions to decimals in order to check a fraction computation, give the class a new fraction computation problem and have everyone use that student's strategy.

And while students won't always completely understand another's explanation, just like we don't always follow others' reasoning completely, it's important that a supportive environment exists where students understand that they are expected to share their thinking and reasoning and to try to understand the thinking and reasoning of others.

57 What about the sideline chatter that sometimes goes on when someone is presenting?

If sideline chatter occurs when someone is presenting, stop the discussion and insist that all students focus their attention on the speaker. That said, there will be times when a presenter's comment sparks a spontaneous reaction and several students break into conversation that is relevant to the point under discussion. Remember that talking can help students interpret an idea and make sense of it. At times like this, you may choose to allow students a few moments to talk to a partner or group and share their thinking about the idea presented. Then call the class back to attention and ask the presenter to continue.

58 During class discussions my students' eyes glaze over at times and I suspect they are not following what is said. What are some ways to deal with this situation?

It's important for teachers to be mindful of how students are receiving what is being presented. If the speaker shares an especially complex but worthy idea and you can see that they're losing the class, stop the speaker and ask students to discuss the idea with a partner or small group. On the other hand, if the speaker is losing the class because what they're sharing is far beyond what most of the class understands, or if it departs from your purpose for the discussion, acknowledge the presenter's contribution and redirect the discussion. Be mindful of when students need a break to process information before the discussion continues.

59 Students who make mistakes during class discussions are sometimes laughed at or taunted. How can I prevent this from happening?

Hold students accountable for behaving according to your expectations, and be respectful of students when they make mistakes. Treat them the same way you expect them to treat their peers when a mistake is made. If you yell at a student or humiliate her in front of the class, you're not modeling respect. Adolescents are very aware of this and humiliating or yelling at a student will likely cost you her cooperation and worse, her respect. At the same time, it's important to consistently address students when they are not following your expectations, because if you just let it go, you're sending a message to students about what is acceptable in your classroom. Correcting students privately gives them an opportunity to save face.

Confront inappropriate behavior directly and explicitly. Your students should know that laughing at or putting down a student who makes a mistake is not tolerated in your classroom. If they believe that a mistake has been made, they can raise their hand calmly and politely explain their thinking. Then the presenter should be given a chance to address the mistake and correct it. Your job as the teacher is to ensure that any mistakes are dealt with respectfully and as rich opportunities for learning.

60 How can I deal with students' mistakes or erroneous ideas during class discussions without making them reluctant to share in the future or, worse yet, turn them off math?

Ideally, you've set up the classroom environment so that students know to respectfully point out mistakes when they occur *(see Question 59)*. But sometimes students will be unaware of an error and, therefore, will not bring it to the attention of the speaker. In these situations you need to step in—you do not want to leave students thinking that their erroneous thinking is valid.

It's important to acknowledge that mistakes are a natural part of learning and should always be viewed as learning opportunities. Handled appropriately, the incorrect or inefficient ideas that are shared can prompt highly useful and rich discussions.

During a discussion, consider the following ways for dealing with erroneous ideas:

- Ask a clarifying question. Sometimes it's enough to ask a question such as *"So are you saying that a negative times a negative is always negative?"*
- Respectfully point out a contradiction to the student's thinking. Adolescents don't want to lose face in front of their peers, so it's important that your tone of voice does not convey challenge. For example, a student might explain that when finding the probability of rolling a sum of 6 with two dice, there are only three ways: 3 + 3, 1 + 5, 4 + 2. To offer a contradiction, you might say, "Here is a pair of dice, one red and one green. If I roll the two dice, I might get a 4 on the green die and a 2 on the red die. That's one way it's possible to roll a sum of 6. But if the 4 is on the red die and the 2 is on the green die, that's another physically different way to roll a sum of 6, so I must count it." Then give the student a chance to revise the answer.
- Give the student the opportunity to rethink their contribution. Say, *"I'm not sure about that. Can you explain it again?"* Often students will self-correct when given another opportunity.
- Give the student options for what to do next. Say, *"I noticed that other students got a different answer. Would you like to hear someone else's idea or give it another try?"*
- Explain that the student's answer is correct for a different problem. Tell the student, *"Your answer of two hundred and twenty-five would be correct if I asked you to*

multiply fifteen times fifteen, but the problem was to multiply fifteen times sixteen." Give the student the opportunity to try again.

61 **There are times when I make a mistake in front of the class, miscalculating or stating an idea incorrectly. What's the best way for me to deal with this situation during class discussions? Could I use these occasions as a way to model for students how to handle making mistakes during class discussions?**

Every teacher makes mistakes. When you make a mistake, handle it just like you want students to respond when they make a mistake. If you catch your own mistake, say, *"I just realized that I made a mistake. Did anyone notice what mistake I made and would you please correct it?"* This encourages students to listen carefully, and helps them cement their own thinking when they verbalize it to the class.

If you weren't aware that you made a mistake and a student points it out, thank them and restate their correction. If students begin to laugh at you or tease you, remind students that you are also a member of the learning community and that you expect to be treated in the same respectful manner as they treat one another when mistakes are made. You may want to point out that you don't laugh at them when they make an error and that you expect them to extend to you the same courtesy.

| Six | Incorporating Writing into Math Class |

Ken sat down at his desk to review a stack of papers from his eighth-grade math class. The day before, his students had solved a complex problem involving exponential growth. During class, Ken had circulated among the students as they tackled the problem, and he was impressed with their solution strategies. For homework, they were to explain in writing how they solved the problem, and Ken was looking forward to reading their responses. As he went through their written work, however, he found that it didn't reflect the thinking his students had done in class. Students wrote about how they enjoyed solving the problem, but they didn't say very much about their mathematical thinking. Ken wondered how he could approach such assignments in the future to better elicit from students the kind of responses that could further and cement their understanding.

As he was thinking about this issue, his colleagues Dan, Eli, and Cathie stopped by Ken's classroom. After Ken expressed his disappointment with the writing assignment, Dan asked, "I'm curious about why you gave the assignment in the first place. Isn't it enough that the students were successful in finding the right answer?" Dan hesitated for a moment and then continued. "Okay, I've heard that we should have students write in math class. But I'm still not sure why it's so important."

Cathie spoke up. "I think writing can help students get a better handle on math, but I need help figuring out what sorts of writing assignments are worthwhile. My students have difficulty reading their work with a critical eye, and I'd like to know what kind of feedback is useful."

"I'd be happy to discuss these issues with all of you," Eli said, "if you can help me figure out how to deal with the stacks of papers I now have from asking my students to write!"

These comments highlight some of the concerns that math teachers grapple with when they attempt to incorporate writing into their classrooms. Math has traditionally been viewed as a subject that uses symbols rather than words. But the process of writing requires gathering, organizing, and recording one's thoughts—tasks that are central to doing mathematics.

62 Why is it important to have students communicate in writing in math class?

Talking and writing helps students make sense of mathematics and think more clearly and deeply about what they are learning. When students write, they have to organize and reflect on their thinking and revisit their ideas. Writing provides an opportunity for students to reinforce and extend their thinking. Furthermore, it helps teachers understand what their students are learning, thus informing their teaching practice. In recent years, communication in mathematics education has become increasingly important. The communication standard in *Principles and Standards for School Mathematics* (National Council of Teachers of Mathematics 2000c) notes that "instructional programs . . . should enable all students to

1. organize and consolidate their mathematical thinking through communication;
2. communicate their mathematical thinking coherently and clearly to peers, teachers, and others;
3. analyze and evaluate the mathematical thinking and strategies of others;
4. use the language of mathematics to express mathematical ideas precisely" (268).

63 What ingredients are necessary to support adolescents' writing in math class?

First, a classroom environment that encourages and supports communication and risk taking is important. Many adolescents want to be like their peers, so fostering a classroom culture in which sharing ideas without fear of being singled out, teased, or put down is necessary.

Second, students need something worthwhile to write about. Thus as you plan for instruction, select interesting and complex tasks that promote important communication. A worksheet with dozens of multiplication problems on it is not likely to provide students with a worthy reason to write. Third, clearly establish the audience for their writing. Students need to know whom the work is for. Sometimes the audience is you, their teacher. Other times they need to know that their work might be seen by other teachers in the school, by their classmates, shared at conferences, shown to their parents, and so forth. Fourth, to "prime the pump" before writing, hold frequent discussions in which students have "opportunities to

• think through problems;
• formulate explanations;
• try out new vocabulary or notation;
• experiment with forms of argumentation;

- justify conjectures;
- critique justifications;
- reflect on their own understanding and on the ideas of others." (National Council of Teachers of Mathematics 2000c, 272)

It's a good idea to have students talk in pairs or small groups before whole-class discussions, so that students are given the opportunity to check out their thinking with one another before making it "public." When students talk with one another, they can formulate their thinking and ideas *before* they write. *(For help with leading class discussions, see Chapter 5.)* Furthermore, the more specific the assignment students are given when first learning to write in math class, the better. It's easier to describe your strategy for a game, write a conclusion about a data set, or explain a computation procedure than it is to describe your thought process in developing that strategy, devising that conclusion, or determining the computation procedure you needed to employ. Finally, a school culture in which students are expected to write in all subject areas supports students in developing proficiency in writing across these subject areas.

64 What types of writing assignments are appropriate?

There are many worthwhile writing assignments to give in math class. Regardless of what you assign, however, think about your purpose in assigning it. Writing is useful when students solve math problems, because students can use their writing as a way of monitoring and thinking about their problem-solving processes. There will be times when you want students to describe, justify, verify, and extend or generalize their mathematical thinking after solving a problem. Having them do so in writing enables them to engage the problem in a way that can reinforce their understanding—or highlight their difficulties with an inquiry. When you want students to reflect on their learning, have them write about their developing understanding and what questions they still have. Journals or logs are good places to record these kinds of thoughts. Sometimes you'll ask students to write about a specific math concept or big idea; you might give them a prompt such as *What is algebra?* When you want students to record information from a discussion for later use, have them keep a discussion notebook or develop a glossary of terms that come up in discussion. If your students have collected and displayed data in some form, encourage them to write about their observations about the data. If students have engaged in an activity or game that requires the use of logical reasoning, have them record their strategies. As Marilyn Burns writes, "Incorporating a mix of writing activities gives me broader insights into my students' math experiences" (2004, 33).

65 I'm intrigued by the idea of having students write as they solve problems. What might this look like in my math classroom?

For an inquiry involving experimental and theoretical probability, I gave students, working in pairs, two spinners:

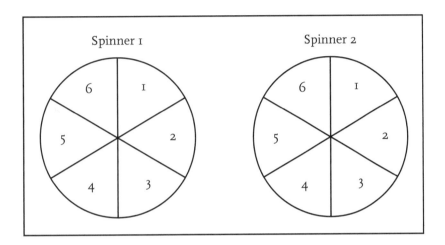

One student would be Player A and the second would be Player B. For each turn, the partners would spin their spinners and add together the numbers spun. Player A would score a point if the sum was even. Player B would score a point if the sum was odd. The object of the inquiry was to figure out if the game was fair—that is, did each player have an equal chance to win? If so, why? If not, how could they make the game fair? Students were asked to record their findings in writing.

Students worked animatedly, spinning the spinners and discussing the problem and their results with enthusiasm. They would talk with one another, write for a bit, talk some more, refer to something they'd written in order to make a point, and write some more. As they worked, I circulated around the room and read their responses. Gianfranco wrote, "Yes! They both have an equal chance to win because there is the same number of even numbers on one spinner as odd." Had I just asked students to state whether the game was fair, Gianfranco's response would have been correct, but his explanation was underdeveloped for me to know if his reasoning was correct.

Tony wrote that he thought that the game was not fair. He incorrectly listed all of the possible pairs of numbers you could spin with the two spinners:

1-1	2-1	3-3	4-4	5-5	6-6
1-2	2-3	3-4	4-5	5-6	
1-3	2-4	3-5	4-6		
1-4	2-5	3-6			
1-5	2-6				
1-6					

In coming to a conclusion, Tony wrote, "The probability of getting an even sum is $\frac{12}{21}$ for Player A and $\frac{9}{21}$ for Player B. So player A has a higher probability of getting a point."

May, however, reasoned correctly by explaining that there were thirty-six possible sums, not twenty-one. She wrote the following:

1,6 = O	2,3 = O	3,4 = O	4,5 = O	5,6 = O	6,1 = O
1,5 = E	2,4 = E	3,5 = E	4,6 = E	5,1 = E	6,2 = E
1,4 = O	2,5 = O	3,6 = O	4,1 = O	5,2 = O	6,3 = O
1,3 = E	2,6 = E	3,1 = E	4,2 = E	5,3 = E	6,4 = E
1,2 = O	2,1 = O	3,2 = O	4,3 = O	5,4 = O	6,5 = O
1,1 = E	2,2 = E	3,3 = E	4,4 = E	5,5 = E	6,6 = E

May concluded on her paper that "Player A and Player B have an equal chance of winning because there are 18 out of 36 ways for Player A to win a point and the same for Player B to win a point, because there are $\frac{18}{36}$ even sums and $\frac{18}{36}$ odd sums."

Asking students to write provided me with a window into their thinking, and their ideas and reasoning were useful as I planned subsequent instruction.

66 *I think it's important for students to describe, justify, generalize, and verify their mathematical thinking. But does this mean that students always need to write long explanations with complete sentences?*

No. Explain to students that it's important that they include details and explain their thinking as thoroughly as they can. But they don't need to write a novel! Including words, numbers and diagrams, charts, and symbols are all useful ways of representing thinking.

My colleague, Nicole, was teaching a probability unit to her students. One of the problems in the unit was *The Integer Dice*:

Travis and Nathan were trying to be clever and invent a new dice game to fool a friend. They made two dice with the following six numbers on it: −3, −2, −1, 0, 1, 2. Before they could invent a game that ensured that they would win, they did some problem solving to determine the probabilities with the dice. If you roll the two dice 100 times, what sum will occur most often? Explain your thinking and your answer completely.

Students worked together to determine the answer, and shared their findings during a class discussion. Students were asked to explain their reasoning and answers in writing. Ted used a combination of charts, lists, and words to clearly explain and verify his thinking. (See Figure 6–1.) Sharquela took the assignment a step further and

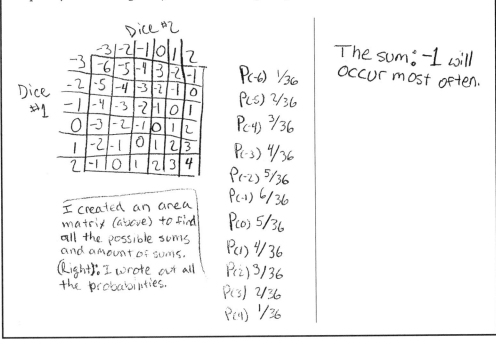

The Integer Dice

Travis and Nathan were trying to be clever and invent a new dice game to fool a friend. They made two dice with the following six numbers on it: −3, −2, −1, 0, 1, 2. Before they could invent a game that ensured that they would win, they did some problem solving to determine the probabilities with the dice. If you roll the two dice 100 times, what sum will occur most often?

Explain your thinking and your answer completely.

Figure 6–1 Ted's solution incorporated writing.

Continued on page 88

Verification

roll #1	roll #2	sum
-3	-3	-6
-3	-2	-5
-3	-1	-4
-3	0	-3
-3	1	-2
-3	2	-1
-2	-3	-5
-2	-2	-4
-2	-1	-3
-2	0	-2
-2	1	-1
-2	2	0
-1	-3	-4
-1	-2	-3
-1	-1	-2
-1	0	-1
-1	1	0
-1	2	1
0	-3	-3
0	-2	-2
0	-1	-1
0	0	0
0	1	1
0	2	2
1	-3	-2
1	-2	-1
1	-1	0
1	0	1
1	1	2
1	2	3
2	-3	-1
2	-2	0
2	-1	1
2	0	2
2	1	3
2	2	4

$P(-6)\ 1/36$

$P(-5)\ 2/36$

$P(-4)\ 3/36$

$P(-3)\ 4/36$

$P(-2)\ 5/36$

$P(-1)\ 6/36$

$P(0)\ 5/36$

$P(1)\ 4/36$

$P(2)\ 3/36$

$P(3)\ 2/36$

$P(4)\ 1/36$

(Far left): I created a systimatic list showing all possible sums and amount of sums.

(Left): I wrote out all the probabilities.

The sum that will occur most often: -1

Figure 6-1 (*continued*) Ted's solution incorporated writing.

described a game using the dice in which one player would have better odds of winning. (See Figure 6–2.) Their writing reflects their approaches to solving the problem. An assignment that just asked for the answer would not have accessed these students' rich thinking.

The Integer Dice

Travis and Nathan were trying to be clever and invent a new dice game to fool a friend. They made two dice with the following six numbers on it: −3, −2, −1, 0, 1, 2. Before they could invent a game that ensured that they would win, they did some problem solving to determine the probabilities with the dice. If you roll the two dice 100 times, what sum will occur most often?

Explain your thinking and your answer completely.

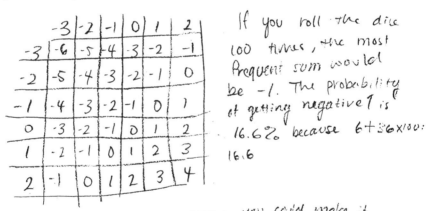

If you roll the dice 100 times, the most frequent sum would be −1. The probability of getting negative 1 is 16.6% because 6+36×100= 16.6

If you wanted to win the game you could make it so that you were Player A and they get the same amount of points for every negative number rolled. So if a −6 was rolled, you get 1 point. Player B would get 1 point for every positive number rolled. This would be unfair because the probability of rolling a negative is 21/36 which is 58.3%. The probability of a positive is 41.6%.

Figure 6–2 Sharquela's solution incorporated writing.   *Continued on page 90*

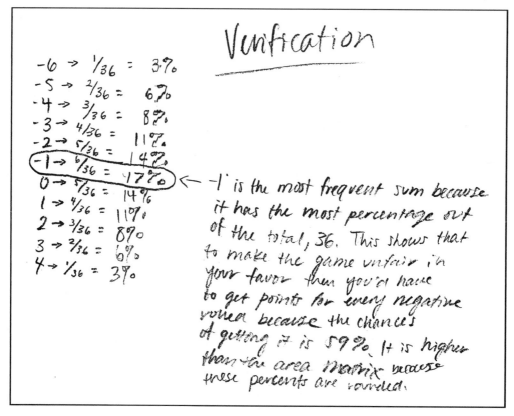

Figure 6–2 (*continued*) Sharquela's solution incorporated writing.

67 What help can I give my students in encouraging them to keep meaningful notes?

Many teachers ask their students to copy things from the overhead or board. As adults we understand the uses of note taking. But not all adolescents are clear about why they should do this and how it might be useful. Many are not sure how to take notes, and for some students, such as English-language learners (ELL) or special education students, note taking can be challenging. (Ask ELL and special education teachers about how to support students who are learning English or who have an Individualized Education Program [IEP]. Their suggestions and tips are likely to benefit everyone.) If you encourage students to use their notes during class assignments and on assessments such as quizzes and tests, they will be more motivated to keep useful records of discussions and presentations. Regardless of when students use their notes, however, they should be encouraged to write only what makes sense to them.

For example, Dalton's students were about to start learning about linear functions. He began the unit by showing a transparency of a ten-by-ten grid with the border shaded:

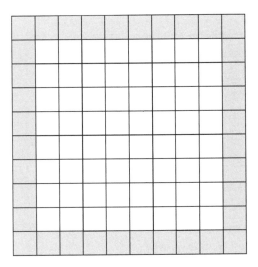

He asked the students to work independently to figure out how many squares are in the border without counting one by one. After a moment, students were prompted to explain their methods for finding the answer. Dalton recorded their methods numerically on the board, labeling each method with the name of the student who offered it. To see if the class could interpret others' methods and apply them in a new situation, he put a five-by-five grid on the overhead with the border shaded in and asked the class to figure this grid's border using two of the ways that had been offered by students in figuring out the ten-by-ten border. He continued the investigation with a "secret grid," telling the class that he had another square grid hidden in his pocket that he'd eventually show them. Dalton asked the students to write a description of how they might go about figuring the squares in its border to support students in thinking about how to generalize. He then coached the group in connecting to algebra symbols the words they had used in describing their methods, and he had them represent their original method algebraically. Finally, Dalton revealed the secret grid and the class determined the number of shaded squares on its border. As the lesson unfolded over two days, many students took notes.

Dalton embedded writing into the lesson in two important ways. First, students had to make sense of the relationship between their arithmetic method for computing the shaded border squares and their written description. Second, they kept notes that illustrated multiple representations of linear functions in meaningful ways for future reference. (See Figure 6–3.)

The Border Problem part 2

Ki's rule: To find the number of squares on the border, I add the number twice (The number of squares on the side) and the take the number and subtract two and add that number twice, last I added it all together.

Example: #4

4+4=8 2+2=4 8+4=12 sq.

S=# of sq on a side

B=# of sq. in the border

$4+4+(4-2)+(4-2)=B$
$\;(S)\;(S)\;(S)\qquad\;\;(S)$

Jerry: Take the number subtract 1 and multiply it by 4 (4=# of sides)
$B=4(S-1)$

Cal: Multiply the number by 4 (# of sides) and subtract 4
$4S-4=B$

La'tiya Subtract 1 from the number and multiply it by four and add 4.
$4(S-1)+4=B$

Eugene Multiply the number of squares on a side by the same number to get the total number of squares (area), take the number of squares on a side and subtract 2 and multiply that number by two (this is the total number of squares inside of the border). Then subtract the number of squares inside of the border from the total number of squares within the whole square and you get the number of squares in the

Figure 6–3 Minh's notes on *The Border Problem*.

68 I understand why it's important to have students write in math class. But how do I get my students to explain their ideas and thinking in writing?

Explain to students at the beginning of the year that part of their responsibility in math class is to write about their thinking, both to aid them in their learning and to help you as their teacher plan effective lessons for them. I have found that it helps to acknowledge that some students love to write, some hate it, and many students are in the middle. Nonetheless, I reassure students that although the writing process takes time, I'll support them in their attempts to communicate their thinking, reasoning, ways of solving problems, reflections on their work, and how they're feeling about what they're learning.

To get started, give a writing assignment immediately after a class discussion. I tell students to write as though they are talking to a partner or group about their ideas. Use writing prompts to help students get started, such as "I think the answer is ____. I think this because . . ." or "What I know about ____ so far is ____."

Some teachers say that their secret to getting students to write is to "nag, nag, nag!" They're kidding, of course, but there is something to be said for continuous but cheerful and kind prodding. Just like learning anything new, writing requires practice. The more often students are asked to write, the better they get at it.

69 What are ways to help students who struggle with writing?

Sometimes students struggle with writing because they aren't sure what they should put on their paper. Special education students and English-language learners frequently have difficulty with writing, regardless of whether they're being asked to write in math class or science class or language arts. The special education or ELL teacher can be a great help in these situations. Other times, however, students have difficulty with writing because they don't understand the mathematics in the problem. So my first reaction when students say they are having trouble writing is to try to figure out where the difficulty lies.

For students who seem to understand the problem but are not sure how to put their thoughts in writing, it's a good idea to prompt them with questions such as "What did you do first?" Wait for their response and reply, "Okay, let's start by writing down what you just said. What did you do next? Why?" Again, wait for a reply and encourage students to write that down next. If they can't tell you what they did first to solve the problem, I ask students something along the lines of "Tell me what the problem is about" and "How might you go about getting started?" to ascertain what they understand, and go from there.

If a student understands the problem but is having difficulty putting thoughts to paper, it also helps to engage the help of classmates. Ask a partner or tablemate to explain the problem, so the student who is struggling with writing can hear one possible solution articulated. Then have the student who is having difficulty writing try again. You'll likely find that as students get better at explaining, they often get better at writing.

70 I have a number of students who are just learning English, and they struggle with the writing I assign in math class. How can I best support them?

There are some general guidelines for supporting nonnative English speakers in math classrooms incorporating writing. First, as you model mathematical talk with the class, be aware that many words in English have multiple meanings, which is confusing for nonnative speakers. "Sum," "table," "rational," and "similar" are examples of words that mean one thing in a mathematical context and something very different in everyday usage. Thus, it's useful to develop language supports, such as vocabulary banks, with students that they can draw on when writing. For a given lesson, identify the key math vocabulary words that English learners will need to know. Consider using a graphic organizer to help students make sense of those words before you pose a problem or investigation:

Word	Definition and/or Word in Home Language	Picture	Example

Have students write the word in English in the first column. In the second column, they are asked to record an accurate definition that makes sense to them or they write the word in their home language. There are many dictionaries online that will provide the word you're looking for in another language. Some such dictionaries are even math specific. (Try using the search term "English to Spanish [or other language] math dictionary." There are several.) In the third column, students draw a picture that illustrates the word, and in the last column, they record an example. Have students then refer to their word bank as they write.

For students who are fluent readers, reading a problem written in English is not likely to present problems. But English learners will need some support in reading the text or problem. Some teachers photocopy the text page and highlight key words for students. Other teachers rewrite the problem in simpler language. While this requires a bit more preparation time, the payoffs with respect to students' comprehension are huge.

Another tip is to carefully consider the context in which you frame mathematical problems or investigations. For example, if you live in Florida and give students a problem that is outside their frame of reference (for example, figuring the volume of maple sugar that can be made from a given amount of sap), English-language learners may experience more difficulty than if you'd ask them a similar problem using a familiar context. Also, be conscious of the speed with which you speak. Slowing down and using gestures helps all of your students, not only those who are learning English.

With respect to writing, it is especially important for students who are learning English to have many opportunities for productive mathematical conversation *before* they are asked to write. During such conversations, have students restate each other's ideas to support their learning and their communication skills. Your job in this case is to facilitate the discourse to make sure it is constructively focused on the mathematics. If English learners get stuck while writing, ask students questions such as "What did you do first? Next? Then what?" and "How did you figure it out?" If you don't speak the English learner's language, have other students who do speak the language translate into English for you so you can make suggestions for putting the ideas on paper.

What's the role of revision? Sometimes my students' ideas are wonderful but there are many spelling and grammar mistakes in their writing. I don't want to 71 make students revise and edit every paper, but I do want them to understand the importance of good presentation. How do I strike the right balance between great ideas and good form?

I let the purpose of the assignment dictate whether or not I expect students to revise and edit their work. If students are writing an "on-demand" papers (in other words, work that you expect them to complete in one class period), I do not expect them to revise their work. I do expect students to revise work if they are working on a long-term assignment (for example, a teacher might assign a *Problem of the Week*, which students are expected to work on throughout the week, and receive feedback from their peers and sometimes the teacher before it is due). In this case, I think carefully about the amount of time students are given to complete, revise, and edit their work when I consider my expectations for the work. If students are working their way into a new mathematical idea, I generally don't ask them to revise every piece of work. If the paper is not clear, however, or if it will become public in some way—posted on a bulletin board, sent home, or presented during a student-led portfolio conference, for example—then I do expect students to revise their written work with a critical eye.

One way to do this is to have samples of writing from students you taught in previous years (be sure to check to see what kind of permission your school or district requires before doing this, and always remove the student's name from the work) to show students. Choose two or three papers that are thorough, two or three others that need revision in some way, and two or three that are bare beginnings. If you don't have papers like these, consider making "composite" examples from a number of papers that you read. It's not as important that the papers are one student's work exactly as it is that the examples allow you to highlight the level of presentation. Discuss with the class what makes a good math paper. Consider developing criteria together. Then distribute the papers in sets to the class, but don't tell them what category you would put them into. Ask students in partners or groups to sort them into three categories: those that meet or exceed the criteria that you set, those papers that are close to meeting, and those that are just beginnings. You might use the following questions as prompts:

- Was the explanation understandable?
- If so, what made it clear?
- What more do you need to know?

Have students talk with one another about their decisions and lead a class discussion about their choices. You want the class to agree about which papers fit in each category.

Next, model ways to "repair" papers when explanations aren't clear or understandable, and then have students in pairs or groups suggest and revise samples of work that need revision. Start with those that need just a bit of revising before asking students to tackle those that need a lot of work. It's extremely helpful to use problems that students have solved and discussed before so they can attend only to the writing instead of both the mathematics and the writing. This helps them when they are doing their own writing and revising.

Jill has her students write specific comments about each sample paper on a sticky note. She explains that it helps students identify what about each paper is clear and understandable and what is not. Jill teaches her students to avoid comments such as "Good job!" because she says that kind of feedback is not very helpful. Instead, state what it is exactly that is good about the job. She sometimes collects and reads the comments on the sticky notes ahead of time to make sure they are appropriate, but more often she circulates around the class as students are writing their feedback to make sure the comments make sense and are not inappropriate. Students use the technique of giving feedback on sticky notes (rather than writing on someone else's paper) throughout the year. Revision then becomes a regular and expected part of her instructional program.

Regardless of the purpose and audience, have students share their writing with a partner or small group before handing it in. Not only is this process helpful for the writer, who gets feedback about his or her writing, but the partner or group benefits from hearing different perspectives and ways to express ideas since the learner has to think about the ideas being presented to understand what's being said.

72 How can writing be used as a vehicle to help my students reflect on their learning?

Writing in math class can take many forms. There are various modes of recording that encourage students to reflect on what they're learning.

Journals or Logs

One way is to ask students to write in a journal or log about the following:

- What you did in class and what you learned
- Questions you have about an idea, a concept, or a skill
- What new ideas you have or what you are wondering about now
- If you worked with a partner or in a group, how things went and what you each contributed that was either "smart" for the task or helped to move the group along to get the job done
- What you found to be easy and what was challenging
- How what you just did or learned relates to something previously learned
- Advice for the teacher about specific activities or problems to use again with next year's students, each with a rationale

Many teachers have students keep such reflections in a journal, spiral notebook, on loose-leaf paper that can be filed in a folder for later reference, or in a log that stays in the classroom. *(See Question 80 for further information on storing students' writing.)*

Portfolios

Another idea is to have students go through each of their assignments and think about what they have learned at the end of a unit or chapter. They might select a variety of pieces that illustrate specific learning targets or other criteria. Ask students to consider a broader audience than just the teacher for their work—such as their parents, other students, other teachers—and then to write a letter to their "interested reader." When students are preparing work for an audience, for example when this work will be made "public" in some way, it's appropriate to take them through the process of revising and editing their written work.

For example, after completing a unit on exponential growth and decay, Jesse wrote and then revised, with help from a classmate, a letter introducing his portfolio work:

Dear Portfolio Reader,

In the exponential growth and decay unit, I learned about exponents, and that $y = ab^x$. I learned how to find the base number by finding the ratio between two numbers and that "a" is the starting number. Another thing I learned during this unit is the workings of a graphing calculator—how to graph, type in equations, change the view, and so on. I also perfected making equations.

    One example of exponential growth in real life is the growth of the population. Another is cells. They split in half to duplicate. The equation would be $y = 2^x$ where x = the generation.

Use the students' portfolios during parent conferences to help them see first-hand what their child has learned. *(See pages 163–64 for information about parent conferences.)* Portfolios are a lively showcase of what a student knows and can do. (See Figure 6–4.)

Assessment Questions

Still another idea to encourage students to reflect on their learning is to ask them to work in pairs or a small group to generate a question that could be put on an end-of-unit assessment. Students should include the answer, ways the problem could be solved, and why the question should be included on an assessment. To do this successfully requires a great deal of reflection and consolidation of knowledge. Of course, you'll want to use some of the students' questions on the assessment so they know that you really do value their thinking and writing. Oscar and Theo, for example, created the following assessment question at the end of a unit on probability:

Sammy got to go to the ice-cream shop. He gets to have three different flavors on his ice-cream cone. He gets two types of cones, waffle and plain. The type of ice cream he wanted was vanilla, chocolate, and strawberry. The scoops can be placed on the cone in any order. How many possible ice-cream and cone choices could Sammy receive? What is the chance of Sammy choosing a waffle cone with strawberry, chocolate, and vanilla (in that order)? If Sammy came back to the store ten times, what is the chance of Sammy getting a waffle cone with strawberry, chocolate, and vanilla?

Comparing And Scaling

In this report I will tell about 4 different things our class learned about in the unit comparing and scaling. They are (1) five ways to make comparasins, ~~crossed out~~ (2) creating and using a rate table to compare rates, and (3) determining unit rates and using them to compare numbers.

Five ways to compare numbers

The five ways to compare numbers are percents, scaling, differences, ratios, and fractions/decimals. Percents are how much out of a whole if the whole was 100. You divide the part by the whole to find a percent. Then you multiply it by 100. For example, if there was 12 kids in a class and 7 were girls, you divide 7 by 12 which is .58 x 100 = 58%. Scaling is where you make a fraction bigger or smaller. For example, $\frac{1}{2}$ could be scaled up to $\frac{2}{4}$, but it meens the same thing. You can also make it smaller. For example, $\frac{4}{6}$ can be scaled to $\frac{2}{3}$. Differences are how much larger or smaller a number is then another. You find it by subtracting

Figure 6–4 Bella's report, included in her portfolio, showed that she learned a great deal about comparing and scaling quantities. *Continued on pages 100–101*

the larger number by the smaller number. For example, If I had 16 pencils, and Dani had 9 pencils, you would subtract 16 by 9 and you would get that the difference is 7. A ratio is comparing two numbers. You don't have to work to make a ratio, you simply put a "to" between two numbers. For example, if there was four girls and 6 boys it would be 4 to 6. You could also write it 4:6. A fraction is made up of 2 parts: The top, numerator, and the bottom, the denominator. You turn a fraction into a percent by dividing the numerator by the denominator, and then multiplying it by 100. The denominator is the whole or toatle amount, and the numerator is the part. For example, 20 kids in a class and 5 girls would be $\frac{5}{20}$ you reduce it by divide the 5 into the 20. $\frac{1}{4}$.

Using a Rate table to Compare ~ Ratios

You make a rate table by makeing a rectangle and putting as many catagorys as you need in it. For example, if my data was "Jenny has a lemonade mix that makes 6 cups, and she needs 24

Figure 6–4 (*continued*) Bella's report, included in her portfolio, showed that she learned a great deal about comparing and scaling quantities.

cups, you would make a rate table like this!

# of batches →	0	1	2	3	4	
# of cups →	0	6	12	18	24	

↑
4 bathes will make 24 cups.

Unit Rates

A unit rate is how much of something one of the objects will be. For example, if tissue was on sale, $1.99 for 2 boxs or $3.29 for 9 boxes to find the price per box you would divide the amount of money by the object. (In this case the number of tissue boxes)

$1.99 \div 2 = 0.995$ $3.29 \div 9 = .366$ ←

costs more per box so this one is the better deal.

This concludes my checkup on Comparing and Scaling. (see back for best work, pre-assessment, and inv. 6, paper pool.

Figure 6–4 (*continued*) Bella's report, included in her portfolio, showed that she learned a great deal about comparing and scaling quantities.

The boys included solutions for their questions and a rationale for why the questions would be appropriate for inclusion on an end-of-unit assessment.

Evaluative Logs

Some teachers have students keep a mathematical learning log that they are expected to write in each day. The log provides an efficient way for students to record their understanding of a math concept or skill. (See Figures 6–5 and 6–6.)

Demonstration logs allow students to reflect on, at the end of a unit, the learning expectations or targets, and record any assignments that demonstrate their knowledge, skills, reasoning, and insights. (See Figure 6–7.)

Finally, in an assessment log, students reflect on the assessments they completed throughout a chapter or unit and comment on each. (See Figure 6–8.)

Mathematical Learning Log

Monday, September 9

Q. What does the **mode** tell us about a set of **data**?

The mode is the most Freequently used whatever the thing is.

mode

Tuesday, September 10

Q. How are a table of **data**, a **line plot**, and a **bar graph** alike?

A table of data, a line plot and a bar graph are Alike becuase they all organize and display data.

Figure 6–5 The writing in Peter's first mathematical learning log of the year was somewhat terse. He became a more prolific and detailed writer as the year went on.

Q. What does the **median** tell us about a set of **data**?

The median is the numerical value that is in the middle of an ordered set. Half the data are above the ~~median~~ median and half are below.

For Example if the ages of 5 kids were put in order and it looked like this:

10, 11, (11), 12, 13   The median is 11.
  1   2   3   2   1
        middle

The median of September 9-13 is 11.

9, 10, (11), 12, 13 ✓
  1   2   3   2

Q. Why is it helpful to give the **range** when you describe a set of **data**?

It is helpful to give a range because if you give a range ~~from~~ from 3 to 10, you know that all of the numbers are between 3 and 10. Also the range helps you figure out the median and mode, because you know the median and the mode are in between 3 and 10.

For example   # of family members.
The range is from 2 to 4. 2 3 4 4 4
The median is 4.

The mode is 4. The median and mode are between 2 - 4.

Q. Can the **mode** and **median** for a set of **data** be the same? Can they be different? why?

The mode and the median can be the same and they can be different. It can be the same because some data has the median repeated the most, so it is the mode also.

For a example: # of pets.   The median is 2 and
            0, 1, (2), 2, 3   2 is the reaped number
                             so 2 is also the mode.

They can be different because a number can be the middle but it might not be repeat as much as other numbers.

For example: # of pets   The median is 1 but
           0, 0, (1), 2, 3, the number most repeated
                            is 0.

Figure 6–6 Tammy's log showed what she learned about measures of central tendency.

Demonstration Log

Write the name of any assignments that demonstrate your knowledge, skills, reasoning, and aptitude with statistics. Write your effort or academic score, too.

Learning Expectation	Evidence of Learning	Score
1. Use tables, line plots, bar graphs, and coordinate graphs to display data.	Ace questions 8 & 9	⑤
	Problem 1.3 Ace 1~6	Ⓞ
	Problem 1.1	Ⓞ
	Problem 4.1 Chart	none
	Problem 4.1 graph	none
	Problem 5.4	
2. Use measures of the center (mode, mean, and median) and measures of spread (range and intervals in a range) to describe what is typical about data.	Problem 1.4	Ⓞ
	The value of your name	Ⓞ
	Ace questions 1~7	O
	Number sense maker · #4	Ⓞ
3. Explain how the median responds to changes in the amount and size of data.	Problem 1.5 Poster	Ⓞ
4. Tell the difference between categorical and numerical data.	Problem 2.2 A~K	Ⓞ
5. Make stem-and-leaf plots, including back-to-back stem-and-leaf plots, to compare data and find the median, mode, and range.	Ace questions 1 & 2	O
	3.1 Follow-up	none
	Ace question #9	4
6. Write an appropriate scale on the vertical (y-axis) and horizontal (x-axis) of a coordinate graph.	Ace Questions 1 & 2	O
7. Tell the difference between the mean, median, and mode as ways to describe what is typical about a set of data.	Numbers sence maker #3	⊙
8. Explain that the mean is a number that *evens out* or *balances* a distribution of data.	Problem 5.2	⊘

Figure 6–7 Kylie found several assignments that demonstrated her learning.

Assessment Log		
Assessment	Score	Student Comments (What do you know best? What do you need to learn better? How can you learn it?)
Preassessment	none	I got 19/26 only 73%
Check Up 1 or Quiz	5	I had the easiest time showing how to figure out how to find areas of parallelograms, squares, triangles, and circles
Unit Test	4	I think I got so many wrong, because I didn't take my time. I understood all the material, but I may have rushed through it.
Reflection Questions	3	I had the hardest time describing information about triangles

Figure 6–8 Cody made notes about his work habits as well as his understanding of measurement.

73 **My teaching partner had her students write a "math autobiography." She said that she learned a lot about her students by giving that assignment, and it helped set a positive tone for the year. What other kinds of general writing assignments might I consider?**

Some teachers set goals with their class to focus on for a specified amount of time, such as a quarter or semester. For example, Judy set a goal for her class to focus on communication, both oral and written. She wanted her students to learn to explain their thinking aloud and in writing. Each student wrote personal goals related to the class goal, including what the goal meant to them, how they intended to accomplish it, and how they'd know that they'd met their goal. At the end of the semester, Judy's students evaluated how they were doing and in some cases, students revised their goals. Other teachers have students select individual goals for math class, and develop a written plan to address them.

My colleague Tom likes to begin the school year by asking his students to write about what they want people to know about them as a learner of math. He also elicits their thoughts about working with others using the prompt "What are the qualities of a good math partner?"

Many teachers find it useful to ask students to write about something that came up during a class discussion. Students could respond, for example, to the prompts "Write about why Rico and Anne disagreed" or "Explain in your own words why Han's explanation made sense" or "Write what you know so far about finding the slope of a line." Use the work the next day so that students see a clear connection between the assignment and their learning. It helps reinforce the idea that you care about their writing.

You can also ask students to write about their general learning in math class, such as "Which activity did you learn the most from and why?" or "Why do you think . . ." or "What would happen if . . ." or "Write a letter to an absent student [or younger student] and explain . . ."

For example, after several experiences with comparing fractions, I asked my students to write helpful hints for comparing fractions. The class had an interesting discussion about what advice they might give someone who had to compare fractions. After the discussion, students wrote about their ideas. Maria wrote the following list of suggestions:

> Compare numerators when the denominators are the same.
>
> Compare denominators when the numerators are the same.
>
> If the numerator is bigger and the denominator is smaller then the one you are comparing it to, then that is the bigger fraction.
>
> Find common denominators and compare.
>
> Convert the fractions into decimals and compare.

74 What are some ways to help students learn math vocabulary words?

Some experienced teachers advise not giving the definition of a new vocabulary word until students have had experience with it. I recommend eliciting the definition of the word from the students after they've used it and are comfortable with it. In my classroom, after extensive use of a word or term, I asked students to do a "quick-write" about the word, during which they quickly jot their ideas about its meaning and give examples of its uses. After sharing with a partner, students volunteer various definitions while I record them on the board. Next, several students consult one of several math dictionaries I keep in the classroom and we edit our list on the board to two or three accurate, but different, definitions. Students finally select the version that makes the most sense to them and record it in a glossary they keep in their math notebooks. The glossary is a twenty-six-sheet booklet, one page for each letter of the alphabet. One student is then asked to write the vocabulary word on a "word wall" posted in the classroom.

Because there are so many math vocabulary words, it's essential to use the words regularly in context when talking with students about them. After doing this, consider taking a word—for example, "variable"—and asking students to write about and then discuss with one another all the things they know about what the word refers to and its mathematical uses.

75 I've been pushing my students to write more thorough explanations when they are solving problems. What should I look for when I read their work and what kinds of written feedback should I give students about their writing?

There are two general questions to keep in mind as you read students' written work. First, what can you learn about each of your students' mathematical knowledge and understanding? Second, what might students' work imply about the specific lesson you taught and about your teaching practice more generally?

With respect to learning about your individual students, read their work and think about the following:

- Did they find the correct solution?
- Does their reasoning support their solution?
- If applicable, does their work reflect a prediction or estimation of the answer or the magnitude of the answer?
- What questions do you have for the student?

As you look at the work through the lens of evaluating your lesson, consider the following:

- Was the problem accessible for every student?
- Did the lesson or problem challenge those students who needed an additional challenge?
- Were there common misconceptions? If so, how might the lesson be modified to address these misconceptions in the future?
- How might I use the students' responses to further their learning?

If you choose to provide written feedback to students, consider writing your comments on a sticky note or on the back of the work. My students request that I do not use red pen and do not mark all over their papers because it makes them feel like they got a lot wrong—even if they didn't! It's also useful to use sticky notes if you're going to ask students to revise their work because you can remove the note after the work has been revised and resubmitted. That said, one teacher I know uses a highlighter and marks those parts of a student's paper that demonstrate good thinking and communication.

Another teacher partners with her language arts colleague, who reads the papers and scores them from a language arts perspective, leaving the math teacher to comment on or score the mathematics in the work. It is somewhat daunting when you first begin reading students' papers, but I can assure you that the process becomes easier over time. I find that I most look forward to reading student work after class. Their insights and ways of thinking about mathematics are fascinating to me and essential for helping me make decisions about what to do next.

76 Must I respond in writing to each student's paper? I'm concerned about the time commitment required. What are some other ways to communicate to students that I've read and carefully considered their ideas?

With class loads of 120-plus students, do not expect to write individual feedback on each paper. It's not realistic. When your goal is to use the work to help you make the best instructional decisions possible, written feedback is not required. On the other hand, it's appropriate to give feedback as part of assessing individual progress.

Certainly, you'll want to read each paper. In the absence of written feedback, let students know that you have read their papers by commenting on their thinking during class time. Use what you learned by reviewing the papers to guide your instruction; for example, you might say, "When I read your papers last night, I noticed that many of you seemed unsure about the second part of the problem, so today we're going to revisit that." Another idea is to read a few selections from several students' papers to highlight points you want to make with the class. I tend to not name the student whose paper I'm reading from because I don't want to draw attention to individual students. The focus should be on the math concept or problem. I might comment, however, on how the paper helped me make decisions about what next to teach them.

77 I'm tired of hearing my students complain when I ask them to write in math class. Why are they so negative, and how should I respond?

It's not uncommon for adolescents to grumble when they're first given writing assignments in math. Typically their bad attitude indicates that they aren't comfortable with writing. I can still hear one of my students saying to me, "Hey, Mrs. R, this is *math* class, not language arts!" It's important to clearly establish with students the purpose of writing in math class. Explain to them why you're having them write. Give them opportunities to write for an audience and make their work "public." Remind them about the importance of practicing when learning something, like a new sport or musical instrument. Practicing is what helps us improve. Remind them that they'll get better at writing over time and with practice. Then cheerfully and consistently have students write!

78 When I have my students write, some of them finish their work faster than others. How do I deal with students who finish early?

Be clear about what you expect students to do when they finish with their writing. Prepare a list of activities from which students can choose when they are done with their work. Some good activities to include on such a list are games that students know the rules for and can play again and again, and individual puzzles or brainteasers. Students could also begin work on the day's homework assignment.

79 When students work together, will one paper from the two students suffice or should I have each student individually write up the assignment?

I recommend that students write their own paper. Writing helps each student process their thinking. When only one paper is expected per pair or group, I typically hear one student complain that they "got stuck" with the activity because their partner or group refused to help out, or I hear complaints that the writer included only their own ideas, not those of their partner or group members. To avoid these common status issues *(see Chapter 3 for information about dealing with status issues in the classroom)*, I insist that each student contribute their own paper most of the time. The exceptions are when students are producing a product such as a poster or project.

Have each student give their paper to a peer who did not solve the problem with them initially, to get feedback about how they communicated their ideas. Again, have students write their feedback on sticky notes. Once the partners have their papers returned, they should read the feedback together and make revisions as needed. What I do to ensure quality is as follows: when it is time to collect the papers, I take both papers from the partners (or all three or four papers if a group has worked together) and I hold them behind my back. I shuffle the papers, then I hold them in front of me again. I explain to the class that whoever's paper is on top is the paper I will score. Thus there is an incentive for all students to care about what is written in their papers! Such a process happens best in a classroom environment that supports risk taking and accountability for one's work and behavior. *(See Chapter 3, "Creating a Productive Classroom Environment," for more information.)*

80 Where should I store all of my students' writing?

It's a good idea to have students keep their writing and papers to promote responsibility and accountability. Furthermore, doing so gives them access to their earlier work as needed.

Journals

Some teachers have students keep all of their written work in one journal. Composition books with graph paper are especially useful for this purpose. There are several advantages of this system. All of a student's work and thinking is in one place, so their progress can be easily tracked. If a student wants to refer back to work done earlier in a unit or earlier in the year, the appropriate pages are readily available. If students write a table of contents that includes the date, title, and page number on the first page or two then it is easy to find any given assignment. A disadvantage of journals is that there can be a conflict when you want to collect them for review but students need them to do homework. Also, if you want students to keep handouts in their journals, these will need to be cut down to fit in the journal and then taped or glued inside.

File Folders

Some teachers keep a file folder for each student in a file cabinet or box in the classroom that students have easy access to. If you assign one file drawer or box or crate per class, and keep a folder filed alphabetically with each student's name on it, their work can be kept in one place. Not every piece of work should go in the folder, but include those samples of work that give insight into the student's thinking, reasoning, skills, and conjectures. You'll need to work out a system for filing papers, however. Some teachers appoint students as their "teaching assistant" for any given class period; that student's job is to do clerical and other tasks, including filing papers.

An advantage of this system is that since the work that students do is in one place, it's easy to see a student's progress. You don't have to worry about papers getting lost from a binder, and students can periodically review and organize the contents of their folder. Having the work all in one place is useful for when you want students to reflect on their work and thinking at the end of a unit or chapter.

A disadvantage is that if a student wants to look back at a particular piece of work, it means going to the file cabinet and pulling it out, which can be time consuming and chaotic if large numbers of students want to get inside their folder at any time. One way to address this problem is to have a student assistant distribute the folders at the beginning of class and refile them at the end of class. Another disadvantage is if papers are just stuffed into folders as they are returned to students, you'll need to develop a system for keeping the papers organized in the folders. Making a table of contents that is periodically added to is one way around this problem.

Notebooks or Binders

Still other teachers have students keep a notebook with dividers in which to keep written work. Sections can be assigned for various assignments such as vocabulary, class notes, homework, classwork, and so on. Spiral notebooks can be kept inside the binder. An advantage of a notebook is that it's always with students, and they can refer to it easily when they need something. Again, having students create a table of contents is useful for finding things.

A disadvantage of notebooks is that without careful monitoring, they can become overstuffed and disorganized, including papers from other subjects, or "black holes," containing papers that are never seen again!

Some teachers like using a combination of storage methods. You might have students keep a notebook but also file work that has been revised or scored in a file folder that becomes a portfolio of student work. Or you might have students keep a journal that stays in the classroom all the time so it can't be left behind in a locker or "lost." You will probably need to experiment to find out which system makes sense for you and your students.

Seven | Using Manipulative Materials

$\mathbf{T}$he principal of Mountain Middle School called the math team to the office one day after school. "The math materials we ordered have arrived," she explained. "Why don't we open the boxes and figure out how to divide up the materials?"

"This is great!" exclaimed Elizabeth as she dug into a box of brightly colored pattern blocks. "I know my students are going to love using these! They are wonderful for helping students learn about geometry. And they're useful for teaching students about fractions and part–whole relationships. But I wonder about how to help students connect what they're learning with manipulatives to math symbols and equations."

"I worry about that too," replied Cole. "I've also wanted to use manipulatives for scale factor problems. But I've seen too many students handle them inappropriately. I want to think through how to manage the materials so they're taken care of. I know these things are mighty expensive."

"Let's sit down later and talk about how we can help students get started using them," suggested Jane, a new teacher at the school. "I have some questions about how and when to use manipulatives in ways that really challenge students."

Teachers generally believe that manipulative materials can enable students to touch, feel, see, and make sense of mathematical ideas, but they have concerns about their appropriate use. It's important for teachers to carefully think through and discuss with one another the specifics of using manipulatives in the math classroom.

My curriculum requires the use of manipulative materials such as tiles, pattern blocks, and geoboards, but I'm concerned that my middle school students may consider them "babyish." How can I help them understand why manipulative materials are important for helping them learn math?

81

In my experience, most adolescents are interested in getting their hands on manipulative materials. I explain to my classes that during the year, as they learn mathematics, they'll have access to a variety of tools that can help them think mathematically. Sometimes the tools are their classmates or teacher. Sometimes the tools are a book, their notes from class, or other resources. And sometimes the tools are manipulative materials. What's important is that students know that there are many resources in the classroom to help them make sense of mathematical ideas.

I explain to students that manipulative materials are valuable in many ways. They help students make abstract ideas concrete. While pictures and other visuals in books and texts are useful, there's no substitute for firsthand experience, and manipulative materials support their learning by providing them ways to construct physical models of abstract ideas.

Manipulative materials help students think about, make sense of, and express mathematical ideas, giving students concrete examples to help with their verbal and written explanations. They can be used as models to use to find solutions to a wide variety of problems. Manipulative materials help build students' confidence and increase their learning by giving them a way to test and confirm their reasoning. Finally, they make math learning interesting and enjoyable.

It's a good idea to make the first experience with manipulatives something that offers a challenge to students, so that they see the materials as appropriate to their age and math level. For example, Anatoli had planned several activities using Cuisenaire rods to engage his students in thinking about ratio, proportion, and fractions. Cuisenaire rods are blocks that come in different lengths, from one-by-one-by-one centimeter long to one-by-one-by-ten centimeters long. Each length block is a different color:

Length (cm.)	Color (abbreviation)
One	White (w)
Two	Red (r)
Three	Light Green (g)
Four	Purple (p)
Five	Yellow (y)
Six	Dark Green (d)
Seven	Black (k)
Eight	Brown (n)
Nine	Blue (e)
Ten	Orange (o)

Anatoli first had his students look for relationships among the rods, finding all of the rods for which one is half as long as the other. Then students found all the pairs of rods for which one is one-third the length of the other. They did the same for fourths, fifths, sixths, sevenths, and so on. Each time students discovered a relationship between two rods, they recorded it in two ways. For example, when comparing the yellow and orange rods, students wrote:

$$y = \tfrac{1}{2} o$$
$$o = 2 y$$

To extend the experience and create an additional challenge, the next day Anatoli asked his students to investigate the relationship between three particular pairs of rods, and to write two mathematical sentences for each:

1. purple and orange
2. purple and yellow
3. orange and dark green

The subsequent discussion of the sentences students wrote was rich and complex. Ed reported for himself and his partner Eugene: "We wrote $o = 2\tfrac{1}{2} p$. We wrote this because it takes two purples and a red to get an orange rod, and red is half of purple. Our other sentence is $p = \tfrac{2}{5} o$. We got this because when we measure with reds, an orange is five reds and a purple is two reds. So purple is two out of five or two-fifths."

Jenny piped in, "Jim and I did it differently. We wrote $p = \tfrac{4}{10} o$. If you use whites to measure, purple is four whites and orange is ten whites. So it's the same thing, but we just used a different rod to measure with."

It was clear that using manipulative materials had afforded students the opportunity to engage in challenging and complex mathematical relationships.

82 Using manipulative materials sounds intriguing, but I see a potential management nightmare. I can just imagine blocks and rubber bands flying across the classroom the moment they get into students' hands. How can I ensure that students stay on task when using these materials?

Students need to know your expectations for handling manipulative materials responsibly, and then you need to consistently follow though with enforcing the expectations. When introducing math manipulatives—be it tangrams, Cuisenaire rods, or dice—explain to students that during the year they'll be using these materials to help them think about the mathematics they are doing. Tell the class the name of the material. Ask if anyone has used them before. It's likely that some students will be familiar with them. Show where they are kept, and explain how you expect students to retrieve them when they are needed as well as your expectations for their use and how they should put them away. Give students a few minutes to explore the material before you expect them to use it as a learning tool, because students need time to satisfy their curiosity. Do this regularly during the first few months of school.

It's safe to expect that some of your students will misuse the materials initially. Three scenarios come to mind, each with a different method of resolution. The first is the case of the student who is not using the material appropriately for the particular task, yet is not disrupting other students or damaging the materials. A friendly but firm reminder will usually get the student on track. Sometimes a student will choose to misuse the materials in a way that is disruptive or destructive. Despite your best efforts, there will always be students who can't resist creating some mischief with a manipulative—sending a rubber band flying across the room, for example. I believe that by engaging in such behavior, students forfeit their opportunity to work with manipulatives during the class period. I prepare ahead of time work that students can do that involves the same mathematical concepts but does not use manipulative materials. I approach students who are misusing manipulatives and calmly explain to them that these are classroom tools that I expect them to use respectfully and appropriately. Since they have decided not to use them in this manner, they will complete the alternative assignment, and they may not use the manipulative for the rest of the class period. I reassure them that the alternative assignment addresses the same concept, and let the student know that they can try again the following day. It's critical that you don't argue with the student or give in to their pleading or promises that they "won't do it again." Remove the manipulative, give them the alternative assignment, and move on. This sends the message that you mean business, and that you also have confidence that the student will be able to handle the situation more appropriately the following day. In the final scenario, several students disrupt the class by misusing the manipulatives, for example by engaging in a "cube fight." In this instance, stop the entire class, have students put the materials away, and hold a class discussion about the proper use of the materials as well as your expectations for their behavior during math time.

Some teachers introduce one or two materials during the first few weeks of school and work with them until students are consistently handling them appropriately. Only then will they introduce additional manipulatives. Other teachers plan an opening unit at the beginning of the year that introduces many of the manipulative materials that students will use in their class.

Regardless of how you choose to introduce your students to using manipulative materials, it's helpful to give students many opportunities to handle materials properly after you've clearly explained your expectations for their use and care. Consistent follow-through on your part is key to making manipulative materials an indispensable part of your math program.

83 Do students really need time to explore the materials before using them for learning?

It's not enough to simply show students how various materials can be used. Students have to handle the objects and do their own "manipulating." Demonstrations alone are insufficient; imagine holding a snake and stroking its skin in front of the class and expecting students to know how it feels. Hands-on investigation is essential for students in building their understanding.

It's important to give students an opportunity to explore a new material before you ask them to use it for a particular purpose. Students need to satisfy their curiosity when being introduced to a new material. If you don't give them this chance, students will most likely have difficulty settling down to work with the material. In their first experience with a manipulative, let students explore and play: have them line up color tiles and knock them down like dominoes, build towers with Cuisenaire rods, create large mosaics with pattern blocks, and make designs with many rubber bands on a geoboard. After you've provided sufficient time for investigation (and the time needed depends on students' familiarity with the manipulative), ask students to explore the material in a more directed manner. For example, if you are introducing students to the geoboard, consider giving them a set of directions (see page 117) to focus their exploration. One teacher asked her students to work with a partner to find the solutions for a set of directions and to record their solutions on a sheet of geoboard dot paper.

Explorations Using the Geoboard
(Burns 2000, 95)

With a partner, find a solution for problem #1 and record it on geoboard dot paper. Make sure that you number each solution. With your group, check each other's solutions. Compare your results and talk about how your solutions are similar and how they are different. Then go on to number two and do the same.

1. Make a shape that touches seven pegs. Think of the rubber band as a fence and the pegs that it touches as fenceposts.

2. Make a shape that has five pegs inside. Think of the inside pegs as trees inside the fence.

3. Make a shape that has ten pegs outside it that do not touch the rubber band. Think about the ten pegs as trees growing outside the fence.

4. Make a shape that has five fenceposts with three trees inside.

5. Make a shape that has six fenceposts with two trees inside.

6. Are there any combinations of fenceposts and trees that are not possible? Explain.

7. Use two rubber bands. Use each to make a line segment so the two line segments touch a total of nine pegs.

8. Repeat # 7, this time finding a way to make parallel line segments.

9. Repeat # 7, this time finding a way to make the line segments intersecting.

10. Repeat # 7, this time finding a way to make the line segments perpendicular.

11. Make a right triangle with no two sides the same length.

12. Make a four-sided polygon with no parallel sides. Record its name if you know it.

13. Make a four-sided polygon with all sides different lengths. Record its name if you know it.

14. Make a four-sided polygon with no right angles but with opposite sides parallel. What is it called?

15. Make a four-sided polygon that is not a square, not a rectangle, not a parallelogram, and not a trapezoid.

16. Make two shapes that have the same shape but are different sizes and are not squares.

17. Make up your own direction and possible answers for someone else in class. Have them try your direction and record their solution and their name.

84 I'm concerned that my students won't take manipulative materials seriously as tools for learning mathematics.

Manipulatives can help students engage more directly in their learning, but unless they've internalized your expectations for their use, you may find that these materials distract students from the mathematical inquiry. It's important to let students explore the materials on their own for a short period of time, but they must also understand that these objects aren't for play but are an important mathematical tool. Typically students will come to this understanding on their own once they've used manipulatives in a way that supports their learning about a mathematical concept or problem. If you find them getting distracted by the manipulatives, remember to stay calm, even if you are feeling frustrated. Reiterate your expectations for appropriate use of materials. I find that in some cases, it helps to have students cover the manipulatives on their desk with a sheet of paper while I'm giving instructions about an inquiry and I need their full attention.

85 How do I help students develop an understanding of the manipulative material they're using so it can be used as a mathematical tool?

Students need time to build their understanding of manipulative materials, which requires that they use these materials regularly and reflect on what happens when they use them in various ways. There isn't any meaning inherent in the materials themselves; it has to be developed by the learner. Don't assume that the manipulative material immediately has meaning for students, regardless of how obvious it seems to you. When students are building and using fraction kits, for example, it may be obvious to adults how the kits work and the relationships that they can model. But it's not necessarily the case for students. Only when their meaning is established can students use manipulative materials as supports for learning. (Making Sense: Teaching and Learning Mathematics with Understanding *[Hiebert et al. 1997] is a useful resource for learning more about using mathematical tools.)*

86 I know that my less experienced learners will benefit from using manipulative materials, but do my most experienced learners really need them?

Definitely. The challenge of teaching any subject is to find learning activities that are accessible to all learners and, at the same time, are complex enough to engage those who need additional challenge. Manipulative materials are a rich way to do this. For example, sixth graders learning about fractions were given the problem of finding the fractional value of each piece of a tangram if the whole tangram equals one. A few students who finished more quickly were offered an extension: find the value of each piece if the largest triangle is worth one whole. Extensions are useful for making lessons appropriate for students at all levels. It's helpful to think about both scaffolding and extensions before you start the lesson, at least until you have enough experience that you can do this in the moment while the lesson is taking place.

87 How often should students use manipulative materials? Should they be used daily?

Use manipulative materials as often as they fit into the lesson you've planned. Many teachers make the materials readily available for students to use in supporting their learning. Remember, there are a variety of tools that students might use as resources, including grid paper, straightedges, isometric dot paper, class notes, angle rulers or protractors, compasses, scissors, and more. It's important for students to know when it makes sense to use particular materials, and which materials are most appropriate for use with particular problems and explorations.

88 What happens when students don't seem to be able to connect what they are doing with the manipulative materials to the problems they are expected to solve in their text?

In my first year of teaching, I remember being thrilled that students were eagerly using manipulatives in math class and seemed to understand the math they were engaged in, but I noticed that their symbolic work didn't reflect that understanding. An experienced math teacher helped me to realize that students don't automatically make connections between the work that they do with manipulatives and the corresponding abstractions on a textbook page. Sometimes they see these two things as separate. One of the most important challenges in using manipulative materials is to help students make these essential connections. Make a point of routinely asking students to record the numerical expressions or generalizations that emerge when they use manipulatives to help them make sense of a math problem. Have them regularly refer back to the materials to see where the math notation or generalization came from. Help students make explicit the connections between the written math expressions and the manipulative materials. This holds true even for advanced students. It's important that all students are able to tie their learning with manipulatives to visual and numerical representations.

89 What are efficient ways to store manipulative materials in the classroom?

Think through the traffic patterns in your classroom as well as the storage space you have available to you. Make sure students have easy access to shelves or bookcases in which materials are stored. Materials can be stored in bins or boxes. Decide if it makes sense to have tubs of materials available containing enough materials for a table group of three or four students.

Manipulative materials might also be placed in ziptop plastic bags. In this case, choose the heavy-duty kind of bag, and teach students how to close the top completely. You don't want a student picking up a full bag of materials that has not been closed properly and having them spill everywhere! Another tip when using plastic bags is to punch a hole in the upper corner of the bag so air can escape. This saves shelf space when the bags are put away.

Set aside time for students to clean up the materials, and make sure that they are doing the cleanup, not you. Hold them to your high expectations.

Make sure that classroom manipulatives stay in the classroom. You don't want these objects being used inappropriately on the school bus or in another teacher's class. One teacher I know shared the following tip for rubber bands used on a geoboard: give each student only two rubber bands, and have them return them to you on their way out the door to the next class.

Eight

Using Calculators and Computers

The math staff met to talk about the role of technology in teaching math. "I like that my students readily use graphing calculators. They know how to enter and graph equations and have used them to investigate different kinds of functions. I wish graphing calculators had been around when I was in school," commented Sidra.

"Not me," replied Brian, a teacher close to retirement. "They're way too fancy for my tastes. Besides, all I see students do with them is calculate things they should be able to do in their heads. They're just a crutch!"

Brian's comment ignited the conversation around the table. "But our world is increasingly technological," argued Tammy. "I think we need to help students learn when to use calculators and for what purpose. I worry about what will happen if students don't have access to calculators. High school teachers expect them to know how to use them, students are expected to use them on standardized tests, and many jobs require their use."

"I agree," said Doris, "and the same can be said for computers. I just got three computers in my classroom and this year the computer lab is available to us for teaching math. I need to find appropriate ways to use computers in math class."

"Hold on. Let's take these one at a time," suggested Sidra as she wrote the questions on chart paper. "What else are you wondering about? Let's record all of our questions and then figure out how to address them."

Computers and calculators (increasingly called "handheld computers") are helpful tools for learning mathematics. Currently, 95 percent of students in grades six through eight use computers, and 70 percent of those students use the Internet (DeBell 2005). Computers and calculators are here to stay. Thus, like any tool, both teachers and students need to be aware of when and how they are useful. It's important to think through how to best incorporate their use into your instructional program. Some districts have policies about calculator use; it's wise to consult these also.

90 How are calculators and computers changing math instruction?

Think about when you were in sixth, seventh, or eighth grade. Chances are, in English class you used a pencil or pen to write your papers. Perhaps when you had a special project or report, you used a typewriter. Often you consulted a dictionary, thesaurus, or grammar guide. Either way, the tools helped get the job done but didn't substitute for the thinking you needed to do to write a good paper. Today, many students routinely word process their work because the computer makes the mechanical part of the job easier and frees students to focus on what's important: crafting written work that effectively communicates their thinking. The computer also makes revising and editing their work easier.

Similarly, when we were in math class, many of us spent long hours performing cumbersome calculations by hand using multiple sheets of paper. In higher mathematics, we might have used a slide rule. But we no longer use slide rules because calculators and computers do a better job. They don't *do* the thinking for students though. Students still must be able to read and decide what a problem is asking, figure out what mathematics is involved, determine an appropriate strategy or approach, solve the problem, and verify and interpret the result. Calculators and computers can't do all this for students, but they can certainly facilitate the process much like word processing does for writing.

As a result, our math instruction is changing. For example, I used to teach about families of linear functions by asking students to make graphical representations of relationships and observe what happens in each form of representation (graph, table, equation) to the dependent variable as the independent variable changes. Students drew tables, wrote equations, and made coordinate graphs on grid paper to represent various functions such as $y = x$, $y = 2x$, $y = 3x \ldots$; $y = -x$, $y = -2x$, $y = -3x$; and so on. Needless to say, the process was time consuming and tedious and students ran into various difficulties: they misplotted points, mislabeled the scale, and focused so heavily on the mechanics of creating the representations that sometimes they missed the point of the lesson!

Many current curriculum materials routinely include calculator use and I now teach the lesson described above using a graphing calculator, which is a wonderful tool for this lesson (as well as for others, such as investigating the effect of changing the slope or y-intercept with linear functions). My students learned previously how to make a table and enter an equation in the graphing calculator and how to move between both representations. For the lesson on characteristics of function families, I taught students how to use the graphing calculator to make a graph. In the process of using the graphing calculator they learned new skills, such as how to adjust the window settings so that their lines would be visible in the display window and how to read and adjust the scale on the graph. (I found that having students think about the window settings caused them to think about the scale of a graph in ways they didn't when making them with pencil and paper. They could experiment with ideas in ways that were laborious when making graphs by hand. Furthermore, they were much more proficient at considering an appropriate scale for the graph when making one by hand after they had worked with this idea using the calculator.) Students then easily manipulated the independent variable for various functions and made astute observations of how this affected the dependent variable in each of its representations. What struck me was how quickly students mastered the technology and used it to make sense of the concepts I wanted them to understand. Using the graphing calculator reduced the time needed to record the various representations and increased the amount of time students spent developing understanding and reasoning.

Computers have also changed how math is taught. Using the computer, for example, my students use Logo and other dynamic geometry software to experiment with families of geometric objects to focus on transformations instead of drawing them by hand. The software allows students to actually see what happens when they manipulate variables, which provides a powerful visual support to their learning. My students have also used the computer to work with "virtual manipulatives," which are computer simulations of physical manipulatives (such as algebra tiles, geoboards, pattern blocks, tangrams, fraction pieces, pentominoes, and so forth) that allow students to extend physical experiences they've had in math class and interact with the manipulatives to solve problems. Both graphing calculators and computers help to keep the focus on the important mathematics students need to learn and do so in a way that is efficient, effective, and engaging. I can't imagine teaching without them.

91 What are the benefits of using calculators and computers in math class?

In *Principles and Standards for School Mathematics*, the technology principle states that "Technology is essential in teaching and learning mathematics; it influences the mathematics that is taught and enhances students' learning" (National Council of Teachers of Mathematics 2000c, 24). Furthermore, calculators and computers are helpful because they "furnish visual images of mathematical ideas, they facilitate organizing and analyzing data, and they compute efficiently and accurately" (24).

Research indicates that calculators are beneficial to students at all levels, are effective learning tools, and increase students' confidence, persistence, and enthusiasm for mathematics (Pomerantz 1997). Using calculators and computers allows students access to mathematical understanding by helping them focus on the "whys" as well as the "hows" of doing mathematics. In addition, it's been argued that rote computation and monotonous algebraic computations turn many students off from math. Calculators and computers reduce the time needed for tedious computations and allow access to interesting mathematics. Using calculators and computers makes formulating and testing conjectures easier than doing so by hand with paper and pencil, and helps students gain mathematical insight and reasoning. Finally, the more opportunities students have to use calculators and computers, the more familiar and comfortable they become with the tools, thus enjoying a competitive advantage over those students who haven't used these technologies.

92 I know that there are several kinds of calculators available, such as scientific and graphing. How do they differ and what kinds of calculators should adolescents use?

Let's start with the basics. Four-function calculators are simple calculators that allow students to add, subtract, multiply, and divide quantities. Many have a square root function and typically come with a memory key to store calculations as well. They're useful for performing simple calculations and are widely used in elementary schools. Scientific calculators include the four basic functions plus many more preprogrammed functions, including keys that allow the user to calculate squares, square roots, and exponents. Some have a function that converts a fraction to a decimal and vice versa. Graphing calculators can do all of the calculations typically found on four-function and scientific calculators, and are, in effect, small handheld computers with a display window that is large enough to view graphs. Many teachers of fifth, sixth, and seventh graders have their students use scientific calculators, and for eighth graders, graphing calculators. Your curriculum materials can often provide guidance, as can other math teachers in your school.

93 Should students use calculators instead of using paper and pencil and mental math?

No. Calculators and computers complement but don't replace the need for paper and pencil work or the need to perform mental calculations. It's important that adolescents develop the understanding and skills necessary to solve problems and learn when each tool—be it paper and pencil, a manipulative material, a calculator or computer, or their mental math skills—is most appropriate for the job. Keep in mind, though, that unless students know how to use a calculator, that particular tool isn't really an option (as is true for other tools, too.)

One teacher I know writes a few computation exercises on the board and the class discusses how each tool listed above supports finding the answer. During the discussion, most students realize that for some problems such as 200×25, using mental math is more efficient than using a calculator because you can simply multiply 100×25 and double the product to find the answer. For problems such as $\frac{1}{2} \div \frac{3}{8}$, some students prefer to draw an area model to compute the quotient than use mental math or a calculator or computer. For figuring compound interest, however, most students prefer to use a calculator to make quick and accurate work of the calculation. Their teacher helps them verbalize what many students know intuitively—that the tool used to solve any problem should be appropriate for the task.

94 Can calculators really be used to support the development of students' number sense rather than hinder it? It seems like some students use the calculator as a crutch.

Let's tackle the beliefs part of your question first. For some people who did not grow up using calculators and computers in their own math learning, technology is sometimes viewed with skepticism and fear. Some believe that using calculators and computers contributes to mathematical illiteracy, that students will become overly dependent on them and won't learn their basics. But we know that calculators and computers are only as effective as the information entered into them, which requires thinking, reasoning, and problem solving. It's important to remember that in your classroom you probably won't want students to use calculators all of the time. When you want students to solve problems without a calculator, simply restrict their use, but remember to explain why you are doing so. Then focus on the thinking and reasoning needed to calculate efficiently and accurately.

Others believe that math should be hard work, and that using calculators or computers means a student is taking the easy way out or is lazy. It's a good idea to examine such beliefs to see what might be underlying them. Since we want our students to be successful in math, it makes sense to engage students in mathematical thinking in ways that are interesting, challenging, and enjoyable rather than trying to make math difficult and inaccessible. Using calculators and computers in math class are good ways to do the former.

That said, it's important that students learn about the benefits and limitations of technology available to them in the classroom. To address this, my colleague Marilyn has her students as a class generate two lists after working with calculators and computers in her math class: "Advantages of Calculators" (or computers) and "Disadvantages of Calculators" (or computers). The discussion about the lists helps students become more critical consumers of technology and learn to use it in appropriate situations.

Now for the first part of your question, how calculators can be used to support the development of number sense. Say you are interested in helping students make sense of decimals to build a foundation of understanding from which they can learn about repeating decimals, scientific notation, and related topics. You might engage your students in an activity called *Multiplication Puzzlers*. For each problem listed below, the idea is to find the missing number by using a calculator and the strategy of estimating and checking. Students should not solve the problems by using division, but instead see how many guesses each takes them. They should record their guesses as shown. For example, to solve $4 \times \_\_ = 87$, students might start with 23 and adjust. Below is a possible solution record.

$$4 \times \_\_ = 87$$
$$4 \times 23 = 92$$
$$4 \times 22 = 88$$
$$4 \times 21 = 84$$
$$4 \times 21.5 = 86$$
$$4 \times 21.6 = 86.4$$
$$4 \times 21.7 = 86.8$$
$$4 \times 21.8 = 87.2$$
$$4 \times 21.74 = 86.96$$
$$4 \times 21.75 = 87$$

It took nine guesses.

Students might try the following other puzzlers:

$$5 \times \_\_ = 96$$
$$6 \times \_\_ = 106$$
$$4 \times \_\_ = 63$$
$$8 \times \_\_ = 98$$

In this case, the calculator is used as a tool to help students think about quantities and how they relate to the problem at hand. It won't tell them what to enter. Instead, students have to use their number sense to think about what they know about quantities in order to make reasonable estimates. Using the calculator supports the development of students' number sense as they rely on their intuitive reasoning about numbers and operations. The focus on estimation helps students develop this intuition.

Hit the Target is another game that requires the use of a calculator. Like in *Multiplication Puzzlers*, when students estimate, they have opportunities to compare numbers and think about number relationships. Forming an estimate involves mental calculation as a preliminary step. And calculating mentally helps students develop their own strategies for applying operations and helps them think flexibly.

In the game *Hit the Target*, students work with a partner. They multiply numbers to produce a product that falls within a predetermined range. The object of the game is to hit the target range in as few steps as possible. Here are directions for the game:

1. Players choose or you can provide them with a target range (730–760, for example, or a smaller range like 600–605), in keeping with the kinds of numbers students are comfortable with.

2. Player 1 chooses a number between 1 and 100 (for example, 25) and records it.

3. Player 2 chooses another number to multiply the first number by, records it, and performs the calculation mentally. He announces his answer to his partner, who mentally verifies the result and writes it. They use the calculator to check their figuring. If the product doesn't hit the target range, Player 2 goes back to the original number and multiplies it by another number as described above and Player 1 verifies and writes the result.

4. Players repeat step three until the product falls within the predetermined target range.

5. Players repeat the game, this time alternating roles.

For example, if the target range is 3,750–3,999 and the starting number is 50, the moves made could be as follows:

$50 \times 90 = 4{,}500$	*Too high*
$50 \times 65 = 3{,}250$	*Too low*
$50 \times 70 = 3{,}500$	*Closer, but still too low*
$50 \times 75 = 3{,}750$	*Within the target range*

Other ranges to try: 850–855, 175–180, 335–340. Even more challenging: 199–200, 3,000–3,001. An extension is to play the game with division, addition, or subtraction instead of multiplication. Another extension is to use the product—not the original number—as the new starting number and figure what number to multiply it by to hit the target range. This version of the game often involves dividing by decimals to reach the target. Again, a calculator is a useful tool in this version.

In both of these activities, a calculator supports the development of students' number sense, but does not do the thinking for them. Adolescents sometimes make place-value mistakes (getting 100 instead of 10, for example) and students need to be able to estimate the magnitude of the answer to know if the calculator is close. A calculator can help them grapple with this idea in ways that are more efficient and engaging than using paper and pencil. *(See Questions 9, 11, and 12 in Chapter 2, "Planning for Instruction," for more about number sense and the basics.)*

95 How can I help my students become more proficient with their basic facts so that they don't always grab a calculator to compute problems they should be able to figure mentally?

It's important that students are aware of their use and purpose, but calculators are not a substitute for students learning to think about and calculate effectively and efficiently on their own. Adolescents need to continue to develop their number sense and computational fluency, especially with rational numbers. As adults, most of us use calculators when we want to save time and a calculation is more complex than we want to do mentally or with paper and pencil. We also use calculators when accuracy is important. We use them for figuring taxes, for balancing our checkbook, and so on. Similarly, students should learn when to use calculators and when other methods of calculating, such as paper and pencil, estimation, or figuring mentally are more appropriate. When students use a calculator they should be able to draw on their skills of rounding numbers and calculating mentally to get a sense of the approximate size of the answer. Furthermore, they need to have strategies to check and repeat the calculation if they are not sure whether it is right.

If you're assessing your students' ability to calculate or you want them to have practice with computation, then it's fine to restrict their access to calculators. Just be sure to explain why you are restricting access. The key is to think about when you want students to use calculators and when you don't.

One way to provide computation practice is to teach students how to play the *Math Bowling* game. Edward first explained to his class how conventional bowling is played and scored. Then he described how to play the *Math Bowling* game. "You take a die and roll it three times," he told his class. "The three numbers rolled make one ball. You have five minutes to write as many different equations as you can using the three numbers. You can use the three numbers in any order and include any operation and any symbols. You can also use the numbers more than once. You'll have to decide what tool is most useful for any given calculation—a calculator, mental math, or paper and pencil." He put a transparency of a set of bowling pins on the overhead.

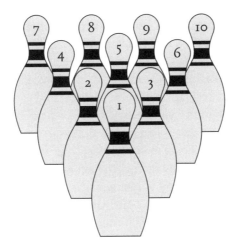

"In bowling," he continued, "you're trying to knock down all of the pins with one ball. In our *Math Bowling* game, the object of the game is to cross out or 'knock down' all the pins with one ball, which is three rolls of the die. Each time you write an equation that yields one of the numbers on a pin, you get to cross off or 'knock down' that pin. So you want to create equations that give as many different answers from one to ten as possible. Let's try playing *Math Bowling* and see what happens."

Edward had a student roll a die three times. The numbers 2, 3, and 1 were rolled. "One equation I can write is twenty-one divided by three equals seven," Edward told the class. So I'll be able to cross off the seven shortly since I've 'knocked it down.' I could also write one times three squared equals nine and then cross off the nine when it's time."

"Oh, I get it," several students exclaimed, and excitedly began writing.

"You can add three plus two plus one and get six. Cross it off!" shouted one student.

"Not so fast," replied Edward with a laugh, pleased by his students' enthusiasm. "Everyone has five minutes to record calculations, and then we'll see how we did as a class. Ready? Go." He set his timer and students quickly got to work. Some used mental math to generate equations, some scrawled equations on a sheet of paper, many used a calculator. After five minutes were up, Edward called time. "Let's see how we did." He reached into a can containing popsicle sticks marked with individual students' names, his method of ensuring that students were called on equitably. The student whose name was drawn offered the equation $12 - 3 = 9$. Edward had the student record the equation on the board and cross out the 9 pin on the overhead transparency. Then he drew another name. That student said, "Three times (two minus one) equals three." She recorded the number sentence on the board and crossed off the 3. "Can I use the example we did earlier? Twenty-one divided by three equals seven?" queried another student. "Sure," Edward replied and had the student cross off the 7 pin on the overhead and write the equation on the board. The next student offered $3^2 - 2^2 + 1 - 1 = 5$ and crossed off the 5 pin. "How about two times (three plus one) equals six?" asked another student. Play continued until all the pins were crossed off.

"That's called a strike," Edward explained. "If we hadn't knocked down all the pins, we would have rolled a second 'ball,' called a 'spare,' and tried again." He went on to explain how bowling is scored and how they would keep track of their progress as a class. He then led the class in a discussion of how the game helps them with their computational skills including practicing their basic facts. He had them identify all the concepts they used in their equations, including exponents, integers, the distributive property, order of operations, brackets and parentheses, and more. Finally they discussed when paper and pencil, mental math, or a calculator was most appropriate. "That was way more fun than just doing worksheets!" exclaimed one student on her way out of class that day. *(Question 145 in Chapter 11, "Handling Homework," gives information about supporting students in memorizing their multiplication facts.)*

96 What are ways to teach students how to use a scientific or graphing calculator?

Just like with manipulatives, providing free exploration time is useful. After the all-important discussion about care of their calculator, distribute the calculators or have students take theirs out. Elicit the basics from them—how to turn the calculator on (and off, if appropriate). Give students time to explore with them. You'll probably notice that some students quickly become calculator "experts" who can readily assist their classmates when needed.

Have students work in pairs and record everything they learn about the calculator. You might suggest the following writing prompts:

To find the square root of a number, _____.

Press _____ to compute exponents.

To convert from a decimal to a fraction, _____.

If you want to make a table of values, _____.

Touch the _____ key to clear the screen.

To graph an equation of a line, _____.

Press _____ to turn off the calculator.

Then have partners share some of their discoveries with the class. Students can publish their results in a "Calculator Guide" or they can be written on posters to hang in the classroom for future reference. Students can add to their guide or poster each time they explore with the calculator, which should be before they use it as a learning tool, especially at the beginning of the year.

A good way to learn about the scientific calculator is to use it to solve an appropriate problem. You might involve students in a number sense investigation called *Getting to One*, which involves decimals. Students begin by entering .5, then +, then =, and = again. After students note that the display shows 1, challenge them to try other starting numbers, press +, and then touch = as many times as needed to try to land exactly on 1 (not all calculators will do this with the keystrokes that I listed, so you'll need to experiment a bit). Encourage students to find as many starting numbers as possible to "get to one" in this way.

Likewise, a good way to learn about the graphing calculator is to use it for a problem. My students had recently completed a lesson designed to introduce them to the idea of variables and how two variables change relative to each other. In the *Jumping Jack* problem, students worked in teams of four. One person was the designated "Jumper," who did jumping jacks for two consecutive minutes. One student kept track of the elapsed time and called out the word "time" every ten seconds for two minutes while the jumper jumped. The third student counted the number of jumping jacks completed by the jumper every ten seconds and announced the cumulative total when the timer called time. The last student consecutively recorded on a table the number of jumping jacks performed every ten seconds. When they were finished collecting data, students made a coordinate graph, recording the dependent variable (jumping jacks) on the y-axis and the independent variable (time) on the x-axis. They explored the relationship between the number of jumping jacks and time. They then investigated what happened when the scale of the graph changed, and made several other coordinate graphs for the jumping jack data using a different scale. The following day, we repeated the activity, but instead of having students record the data by hand, I taught them how to use the table function on a graphing calculator. This time as the "Jumper" jumped, the students entered the number and time data into the calculator. I explained to them how to graph the data using the calculator, and how to change the viewing window and scale. Students were amazed at the power of this tool, and repeatedly asked if they could use it for math problems we did during the rest of the year.

97 My school doesn't provide calculators for students, so they bring their own from home and they're all different kinds and brands. What should I do about dealing with the differences?

One thing to remember is that students today are often much more technologically savvy than many of their teachers! What we think will create problems for students, often doesn't. For example, I had three different brands of graphing calculators in my classroom. We spent a few days learning how to use each kind and then students were given choices about which they preferred. So for any given problem, a variety of brands of calculators were in use in the classroom. Even if two students were using different brands and one person got stuck, their partner was often able to help them figure out what to do.

At the beginning of the year, it can help to have students pair up and explore similarities and differences in how their calculators work. They might investigate the following:

How many digits can be shown on the display?

Do they both have the same keys or are there differences?

How do the arrangements of the keys compare?

Is there a shift key? Is it used the same way on both calculators?

Do they both have memories?

Do the memories work in the same ways?

Try a subtraction problem where the second number is larger than the first; for example, 2 − 9. Do your calculators give the same answer?

Do your calculators give the same answer to 4 ÷ 6?

How do your calculators deal with large numbers? Try entering, 1,000,000,000. What does each calculator display?

How does each calculator handle scientific notation?

Try entering the same number sentence, such as $10 \times 4 + 3^2 − 24$. How do you have to enter the symbols so that each calculator will compute the calculations according to the correct order of operations?

What other differences do you find?

Consider having students change partners and repeating the activity so they can learn about other calculators. Then hold a class discussion so students can share what they learned.

98 **Our school provides calculators for students. A colleague told me that he doesn't make them available to students anymore because they didn't take proper care of them—too many were damaged, destroyed, or stolen. I want my students to have access to calculators. How do I encourage their proper use?**

Calculators—especially graphing calculators—can be expensive. Regardless of cost, calculators, like all classroom matierals, must be handled carefully. When introducing the calculators to the class, demonstrate proper handling. If the calculators have covers, for example, show them how they are to be put on and removed. Explain that such calculators can easily slide out of the covers and onto a table or the floor, so it is important to hold the calculator by its body rather than the cover. Instruct students to gently push the buttons on the calculator using their finger or the eraser part of a pencil. Sometimes students are tempted to use the point of a pen or pencil to push the calculator buttons, and this can damage the buttons.

Teachers must be on the alert for improper use of calculators. One teacher I know had a class set of graphing calculators featuring a clip inside the battery cover that enabled the user to disable the batteries so they appeared to be dead. Several students were using this feature to get fresh batteries from the teacher, and pocketing the other, perfectly good batteries for use in their own personal electronic devices, such as music players. When the teacher caught on, she prohibited them from opening up the battery cover, much less touching the clip inside it.

I have students clear the memory before turning off their calculator at the end of class so that anything entered into the calculator during one class doesn't inadvertently create problems for a student in a subsequent class. (This also addresses improper use of the alpha key, which in my experience students have used to leave messages, sometimes inappropriate, for the next class.) And with graphing calculators, it's essential that the calculator is actually turned off at the end of class to conserve the battery life. Have students do these things routinely, so that they become second nature. If you consistently follow through with your expectations for proper use; there should be few problems with using calculators in the classroom.

99 How do I make sure that my class set of calculators remains in the classroom when the bell rings? I don't distrust my students, but I can see them accidentally walking out of class with a graphing calculator.

One important decision you'll need to make is whether or not to let students take school-provided calculators home. Check with the other teachers in your school to find out if there is a policy or guidelines about doing so. (Some teachers leave a few calculators in the library for checkout.) Regardless of your decision, mark the calculators in some way so that they are easily identifiable as belonging to the school. Many teachers also write a number on each calculator using an engraver or a permanent marking pen. If you assign a specific student a specific number and explain that they are accountable for their calculator when using it in the classroom, it is likely to be well cared for.

Calculators should be kept in a secure location. Some teachers store their calculators in a hanging caddy that has a pocket specifically designed for the purpose of storing calculators. Each pocket has a number on it, making it easy to tell at a glance if all of the calculators are there at the end of class, and which number calculator is missing. Others keep the calculators sequentially stacked in a couple of sturdy boxes or tubs and have one or two trusted students distribute them at the beginning of class. Some teachers prefer to hand out the calculators themselves, and as students leave the classroom at the end of class, they hand their calculator back to their teacher while she stands by the door. Whatever system you use, it's imperative that it's clear and consistently followed. It requires time and energy to do this, but it alleviates the problems that are likely to occur without such a system. If you are lax about calculator care and accountability, your students will be too.

100 Calculators and batteries are expensive! How can I obtain them without spending my own money?

Check with your principal or other math teachers to find out what resources are available to you. If you don't have access to calculators, batteries, and/or some sort of storage system for them, make a list of what you need and approach your principal or math colleague about how to obtain them. In some cases, they may be available from a central storage room or warehouse. Sometimes a teacher may offer to share with you. Sometimes other schools have surplus calculators or batteries you can borrow or have, and as a last resort, they may need to be purchased. Some teachers ask students to bring in a calculator and one package of batteries if appropriate as part of their school supplies at the beginning of the year. Be mindful that this may present a financial hardship for some families, and do not make a student feel lacking if they are unable to comply with the request. Sensitivity is paramount in this situation. There are typically enough students in the class who are able to bring in batteries that battery replacement for the calculators is not a problem.

101 What are some effective uses of computers in math class?

There's a variety of math software available today both for purchase and for free on the Internet that supports and enhances students' mathematical capabilities. Just like you would with books and other print materials, it's necessary to be a critical consumer. Some of the software and Internet resources are rigorous, engaging for students, and reflect what is considered "best practices" in mathematics, while others promote low-level thinking and contain questionable content and methodology. It takes time to sift through what is available, and careful analysis to select appropriate lessons, resources, and software.

Today, many classrooms are networked, allowing students access to the Internet and World Wide Web. The Web is a wonderful source of information. A creative project one of my teaching colleagues assigned to her students was a math World Wide Web scavenger hunt. Most of her students were familiar with accessing the Web, and her questions helped them use the Internet as a rich source of information. She developed a variety of questions and then asked students to search the World Wide Web to find the answers. Students were challenged to "Find two famous athletes who studied mathematics in college"; "Find a country whose land area is smaller than the area of the state of New York"; and answer such questions as "What is a googol?"; and "Who are four important black mathematicians [or women mathematicians, mathematicians from Mexico, etc.]?"

Some teachers post homework assignments and grades online so students and families can access them. Others direct students to "homework help" sites for assistance with out-of-school math assignments.

The Web can also be used to directly support students' learning. Some teachers instruct their students to visit various sites on the Web where they can engage in interactive lessons on any of the five mathematics strands: number, geometry, measurement, algebraic thinking, and statistics and probability. There are sites where students can play mathematical games to support their work in the classroom, solve problems that their teacher or others from around the country (or around the world!) have posted online and then post their solutions and respond to other students' solution strategies. There are Web-based projects that a class can sign up for and participate in. What's significant about the Web is that it is constantly changing and new and interesting ways to use mathematics emerge daily. Many schools and districts now have instructional technology support staff who can assist you in finding your way around the Web. Many middle schools now employ a computer teacher who can also help you get started.

In classrooms that are not networked, some teachers make extensive use of spreadsheets. My colleague has her students play *Spreadsheet Battleship* to familiarize themselves with how to name cells in a spreadsheet. Because a spreadsheet allows students to manipulate variables, constants, and step size, it creates an environment that is ripe for student conjectures, investigations, and "what if?" kinds of questions. Students can use spreadsheets to develop and use algorithms, solve problems, create surveys and questionnaires, administer the surveys, collect data, prepare the spreadsheet and input the data, decide how to display it—by creating bar or double-bar graphs, line plots, circle graphs, etc.—and, finally, analyze and interpret the data, and much, much more.

Other teachers use geometry sketching software and systems-modeling software as well as a plethora of mathematical games to support student learning. Finally, students can find calculators and sometimes even graphing calculators on the computer. Computer software for mathematics and Web-based resources are just waiting for you and your students to discover. Enjoy the journey!

Nine | Assessing and Grading

Susan, Cate, Ruth, and Maggie met to discuss their challenges with assessments. Cate commented, "My students take the required standardized tests each spring, but we don't find out until summer how they did. I use chapter tests and quizzes during the year, but I'm wondering if there are other tools I can use to assess students during a unit to help me make better teaching decisions."

"I want my students to take more responsibility for assessing what they are learning. I'm going to try having them reflect in writing on what they've learned as a way of collecting evidence of their learning and their mathematical disposition," replied Ruth. "It should give me interesting information about what I need to do differently in the classroom."

"I'm thinking about standardized test preparation and ways to incorporate it into students' regular instruction," responded Susan. "I hate feeling like I have to drop everything in March to do three solid weeks of test preparation. I know that standardized tests are not the be-all and end-all, but I still feel it's my responsibility to prepare students for them. There's got to be a better way to help prepare students to take the test without seriously deviating from my regular instruction."

Maggie, a new faculty member, had different concerns. "Standardized test preparation? Self-reflections? I can't think about those things when I'm drowning in paperwork! My grading scheme is so complicated, I'm spending hours on each assignment. Does assessment and grading need to be so complex and time consuming?"

Assessment and grading are two important areas of a teacher's math program. As the comments above illustrate, there are many issues to consider when thinking about assessment and grading. It helps to decide what you want to know and why you want to know it. Keep in mind the following questions: What mathematical goal do I have for students? What is important for students to know and be able to do?

At different times during the course of a unit or chapter, you'll likely want to know about:

- *students' knowledge of math concepts and skills;*
- *how students apply their learning to new, novel, or nonroutine problems or situations;*
- *students' thinking, reasoning, and problem-solving approaches;*
- *students' mathematical disposition.*

Assessments provide ways to gather evidence to inform both teachers and students about these things as well as help students solidify their understanding and identify areas of need.

102 I'm interested in regularly assessing student understanding to help me make better teaching decisions; I don't want to rely solely on the end-of-unit test. How can I make assessment an ongoing part of my teaching practice?

It's helpful to assess students in a variety of ways periodically throughout a unit or chapter. What you learn from such assessments helps you adjust your instruction. Consider having students solve a problem at the beginning of a chapter or unit to help you learn what knowledge they bring with them. Vijay, for example, was preparing to begin a unit on quadratics with his eighth graders. He selected the following problem for his students to solve before he began instruction:

The number sequence 2, 6, 12, 20, 30 . . . follows a pattern. Is 650 in the sequence? If it is, explain which number in the sequence it is. If it's not in the sequence, explain why not.

Another option is to ask students to write about a specific mathematical topic or concept, such as "What is probability?" Their responses will give you information that is often interesting, sometimes surprising, and usually helpful for assessing their understanding and planning subsequent instruction.

Assessing students partway through a unit also yields useful information. Tim was teaching a unit on linear functions to his seventh graders. Partway through the unit, he had them complete an assessment. As he read the students' papers, Tim noticed that about half the students were using integers incorrectly in the context of the algebra unit. His students had engaged in a unit on integers earlier in the year, but based on what he learned from the assessment, he decided to add into the algebra unit some lessons focused on integers. It later proved to be time well spent.

103 A teacher at my school recently attended a workshop on assessment, and came back talking about "summative" and "formative" assessments. What do these terms mean?

When assessment is used to summarize what students have learned, it is often called "summative assessment," or assessment *of* learning. Other times, assessment is used for the purpose of helping students learn. This is often called "formative assessment," or assessment *for* learning, and it helps the student recognize not only what they know and can do but what they need to learn in order to reach a learning target. Formative assessment also informs the teacher about what instruction should come next.

104 What role does teacher observation play in assessment? I often feel like I learn a lot about my students when I observe them as they work on a math problem or share their ideas during a discussion, and I want to formalize these observations.

Observing students as they work and while they share their thinking during class discussions is a powerful assessment tool! The strategies and approaches students use in tackling a problem or explaining their solution strategies to their classmates during a discussion give you important clues about their understanding—or lack of understanding—of the mathematics they are learning. Furthermore, some students are better able to explain their understanding verbally than via paper-and-pencil measures. Shrewd teachers listen to students as they work and interact with others, asking mentally, "What does that indicate about what the student knows, and what, if anything, should I ask or do next?" This is at the heart of assessment.

For example, Antonio's students were working on a unit that focused on computing fractions. He was circulating through the classroom as his students worked on a problem. Antonio stopped by a table where several students were talking animatedly about the problem, and he listened to the group's conversation. Mia was explaining to her tablemates that to add fractions, such as $\frac{1}{8} + \frac{1}{4}$, you just add across the top numbers and then add across the bottom numbers to get the answer $\frac{2}{12}$. Antonio reflected on how last year, he would have noted that Mia didn't understand the algorithm for adding fractions, assigned her some additional problems, and left it at that. But this year, he was curious about her thinking, and while he was tempted to intervene, he waited to hear her tablemate's reactions. "That's not right," replied Daniela. "Think about those fraction kits we made with construction paper. When you lay four fourths next to each other, and put eight eighths underneath them all lined up, and add, you see that one-fourth is the same as two-eighths. So two-eighths and one-eighth is three-eighths, not two twelfths." Daniela quickly sketched the following on her paper:

$\frac{1}{4}$		$\frac{1}{4}$		$\frac{1}{4}$		$\frac{1}{4}$	
$\frac{1}{8}$	$\frac{1}{8}$	$\frac{1}{8}$					

$$\frac{1}{8} + \frac{2}{8} = \frac{3}{8}$$

"Oh, I get it," replied Mia. The girls were ready to move on, but Antonio chose to interrupt them to ask Mia, "Tell us why you thought that you should just add the numerators and the denominators to find the sum."

"Well," she said slowly, "it's like in regular addition you just line the numbers up and add them vertically. So I thought that it would work the same way for fractions even though you're adding across."

"What do you think now?" Antonio probed.

"Well, I think that maybe you have to take the fractions you're adding and line them up like we did with the fraction kit and see how much they equal," Mia replied.

When students are first learning about adding fractions, it's not uncommon for them to believe that adding simply requires adding across the numerators and denominators, and the information Mia provided Antonio gave him clues about the kinds of experiences she might need next in her study of fraction addition.

It's important for teachers to keep a record of the information they get from "listening in" on student discussions. Some teachers use labels preprinted with students' names, which they use to jot down notes about problems or breakthroughs students evidence during group discussions. The teacher then marks the date on the label and sticks it in the student's folder for later reference. Other teachers keep notes on checklists for future use.

105 What are some ways besides multiple-choice tests, which seem to have many disadvantages, of finding out what an individual student knows and can do?

Individual assessments are often used to find out about what a student has learned or needs to focus on. Just as teachers provide a variety of ways for students to engage with mathematics, such as problems, activities, and projects, they should use a variety of assessments. Students can demonstrate their learning through:

- self-assessments, the purpose of which is to reflect on their learning as well as on ideas with which they are still struggling;
- short-answer quizzes or tests, where students solve problems and provide brief explanations, showing their work (Students' understanding of concepts and skills—their mathematical thinking—is revealed when they are asked to explain how they solved a problem. Therefore, asking students to "explain your thinking" or "show all work" in quizzes or tests is useful.);
- a portfolio, which is a collection of student work specifically selected for various purposes *(see Question 109)*;
- projects, which are useful and engaging ways for students to apply what they know; and
- essays and other written communication, to help students learn what they need to know in order to reach a learning target or goal and to help the teacher determine what experiences are needed next.

106 Communication is an important area in math instruction. How can I support my students in becoming more proficient at communicating on assessments what they understand?

If you want students to become proficient on assessments at explaining their thinking and reasoning, they first must have opportunities to explore tasks that require thinking and reasoning. Students learn what you expect by the kinds of assignments you give them. If you mostly assign worksheets or textbook pages that require students to record only the answer, they learn that sharing their reasoning is not what is valued in your class. By contrast, giving students assignments that require them, for example, to share both orally and in writing how they figured out their answer, or how they can justify their solution strategies, helps them learn how to communicate clearly in math class while conveying to them what it is you value in math class.

When students begin explaining their thinking aloud or in writing, they'll need your support. Provide examples of papers that are complete and accurate. Compare these with samples of student work that are not. Encourage students to carefully examine the incomplete or incorrect work and discuss how the responses might be improved. Doing so helps students deepen their understanding of the math involved in a given problem, learn what is expected in a mathematical explanation, and improve their ability to assess and take responsibility for their work. Furthermore, a primary benefit for you as a teacher is that you learn what students do and do not understand, which can help you make decisions about what to do next. It gives you insights into whether students grasped the concept you were teaching rather than just coming up with the correct answer. Many teachers incorporate a student's ability to communicate in writing into their grading system to emphasize how important this skill is. (*See also Chapter 5, "Leading Class Discussions," and Chapter 6, "Incorporating Writing into Math Class."*)

107 I've noticed that my students learn from one another when they work together. As a result, I've sometimes assigned a group project to my students. I'm thinking I can do the same with regular assessments, such as quizzes or tests. What are the benefits of working on an assessment together and what tips can you suggest about determining who should work with whom? What about grading the collaborative work?

Partner or group assessments are inherently formative in nature. They are often used when a teacher wants students to apply their learning to a problem or situation that would be especially complex or difficult if attempted alone. They also provide a way for teachers to assess students' process skills such as communication, reasoning, and so on. Partner or group assessments are beneficial in that students can press one another mathematically as they develop their ideas together. Questioning each other's ideas, or verifying one another's thinking, often requires students to analyze and synthesize ideas, which supports their math learning. Working with others is also motivational because human beings are social and adolescents generally like working with other students. When I asked my students what they think about partner or group assessments, Richard commented that "because I'm working with other people, I don't really feel like I'm being tested and that helps me relax and not worry so much about getting it wrong."

One interesting way to create a partner or group assessment is to have the pair or group come up with three questions and answers that might be put on the end-of-unit assessment. The teacher then can select from the questions students have generated and create an assessment that students now care about. Regardless of whether a particular pair's work has been chosen or not, it takes careful thinking to come up with appropriate questions and answers, thus providing an effective way for students to review what was learned.

There are many ways to group students. Sometimes you might assign partners or groups based on specific criteria. One teacher I know connects English learners with native English speakers, or special education students with non–special education students. Other teachers pair up students who struggle to communicate mathematically with those who are more facile in this area. Sometimes teachers pair up students who tend to get along.

Another way to determine groups is to have students choose. In this situation, students can't blame you if they aren't happy with their choice! It's important to be sensitive to any status issues operating in the classroom, however *(see Chapter 3, "Creating a Productive Classroom Environment," for information on addressing status issues)* and make sure that all students feel valued and included.

Finally, some teachers randomly group their students. It's useful to have conversations with the whole class about what makes a good partner or group. Discussing every member's rights and responsibilities in these situations is time well spent, regardless of whether you make the group assignments, students choose them on their own, or they're randomly assigned.

With respect to grading partner or group work, have both partners record the work they've done on their own papers. *(See Question 79 for more on grading partner work.)* When it comes time for you to collect and score the work, put the two papers behind your back, shuffle them, and see which one comes out on top. That's the paper you score, and both partners receive the same score. This encourages students to help one another revise and edit their written work, which supports their skills in communication.

108 My colleague allows students to retake assessments. Is this practice advisable?

I tend to allow students in my classes to retake assessments. I usually have students retake only the questions they've missed, which has a positive impact on their motivation and their scores. (When their scores increase, so does their motivation, creating a very nice cycle.) Rather than give the same question, I have two or three versions of an assessment prepared so that a student can learn what they missed and try again. For a retake, a teacher friend of mine has his students make up a new question of comparable content and difficulty and answer it. He finds that students have to do quite a lot of thinking in order to create a new question and answer, learning in the process. It's an effective way to combine formative and summative assessment.

109 Should students' assessments be stored, and if so, how?

Assessments provide a record of students' current knowledge and their mathematical growth over time, thus both teachers and students should have access to them. Many teachers have students keep them in file folders or the like in a cabinet or box in the classroom, where they can be easily retrieved and utilized as evidence of learning. Some teachers have students keep a folder of work and assessments for each unit completed during the year. Others have students select work and assessments from an entire year to use as a "demonstration" portfolio. Sometimes portfolios are used as a "celebration" of student learning, in which the most successful assessments and best work are included to share with others. What's useful about portfolios is that they give students the opportunity to reflect back on their learning. They can be shared periodically with students' families during parent-teacher conferences or during student-led conferences. Furthermore, portfolios of work and assessments are useful when students determine learning goals at the start of a new quarter or trimester.

110 I've experimented with a variety of methods for scoring students' work and assessments, and many seem complicated, especially when you have a lot of students. What are some tried-and-true methods for scoring student work and assessments?

Percents, letter grades, narrative comments, points, checks and minuses, rubrics . . . there are many options for grading or scoring student work and assessments. What's important is that the system you choose is clear to both students and their families, while not requiring you to devote every waking hour to it.

Many schools and districts now use report cards that separate students' effort in math class from what they learned academically. Consequently, their teachers grade some assignments for effort only or academics only. It's useful to have two different systems of symbols in this case, to signal to both students and families whether the assignment was scored for effort or academics.

Effort is often used to refer to students' participation in class discussions, persistence, willingness to try, homework completion, and so forth. A seventh-grade teacher, Karma, told me, "I tend to give more effort scores than academic scores because most of the time in the classroom, students are learning new things and practicing them, not demonstrating proficiency. While I do give academic scores for some assignments, such as projects, assessments, or problems that require application of knowledge, perhaps 60 percent of the assignments I give receive effort scores." She uses a ✔+ for work that exceeds expectations for effort, ✔ for work that meets expectations, and a ✔− for work that does not meet expectations. A zero indicates no evidence, such as a missing assignment, for example.

Another teacher uses the following system for scoring effort, which she developed with her students:

A = Almost Always (almost always demonstrates an effort to learn; consistently completes work; is persistent and on task)

O = Often (often demonstrates an effort to learn; usually completes work; is usually persistent and on task)

T = Sometimes (sometimes demonstrates an effort to learn; inconsistently completes work; is sometimes persistent and on task)

S = Seldom (seldom demonstrates an effort to learn; infrequently completes work; is seldom persistent and on task)

When Karma chooses to give an academic score, she uses a rubric or scoring guide to help students know how their work or assessment compares to an academic standard or performance standard. Many districts and states have scoring guides that teachers are required to use. If students help you develop a specific scoring guide, however (see example below), they may be more motivated to attend carefully to the quality of their work since they know what they are trying to achieve.

E = Exceeds (demonstrates and applies strong academic performance that is above grade-level expectations)

M = Meets (demonstrates academic performance that meets grade-level expectations)

C = Close to meeting (demonstrates academic performance that is close to meeting grade-level expectations)

N = Does not yet meet (demonstrates academic performance that is below grade-level expectations)

NE = No evidence (student did not submit enough work to determine a mark for academic performance)

111 Should I grade or score every assignment?

No. Think about when you are learning something new. Do you want to be scored on your practice efforts? Besides, scoring everything you assign will leave you without a social life if you have a large class load because you'll spend all of your free time grading assignments!

112 How should I score homework? I bring home more piles of homework papers than any other kinds, and I know there are more efficient ways to manage the flow than going through a huge stack each evening.

Barbara tends to grade almost everything she assigns from August until October or so, to give students the message that she is serious about their work. Then she eases off of grading every single assignment after students have internalized her expectations. Some ideas she uses for managing the paper flow include:

1. Don't always collect the homework. Look at it during the warm-up or at the beginning of class, then correct it together. Have students make corrections during that time and take notes for future reference.

2. Have some students put answers to the homework on a transparency and present it to the class. Rotate this around the class on different days so that many students are presenting.

3. Check homework only for effort or completion (✔, ✔+, ✔−). Then go over selected answers.

4. Ask the class which homework problems they had trouble with and then only go over those with the whole class.

These techniques should be applied judiciously to avoid spending copious amounts of class time on homework. Consider instead assigning homework that will be used during the next day's lesson. In one class, a teacher assigned *The Perimeter Stays the Same. (See Question 136 in Chapter 11, "Handling Homework.")* The following day, as part of his math lesson, he gave the following instructions for processing the homework:

1. Exchange papers with someone in your group. Check your partner's shapes to be sure that each has a perimeter of thirty centimeters. Also check that the area of each shape is accurate.

2. With your group, examine the shapes and discuss what you notice about the shapes with the same and different areas.

3. Cut out the shape from your group's papers that has the greatest area and the shape that has the least area. Post them on the class chart.

The class then had a whole-group discussion. The teacher asked the students how they made shapes with a perimeter of thirty centimeters and what problems they ran into making the shapes. The teacher then initiated a discussion that focused on what kinds of shapes with a fixed perimeter give the least and most area.

113 How should I prepare students for the tests that the district or state requires?

Find out what assessments your students are required to take. Learn about the structure of those assessments. Find out how closely your instructional materials align with the content of the assessments that so you know what needs to be covered. Talk with teachers from the previous grade level: did they cover everything in the curriculum that they intended to, or were some topics or units skipped? These are some important first steps. Your district curriculum coordinator or testing department should be able to help, as should your principal and math colleagues.

Students are required to take a variety of math tests. Some include only multiple-choice items; some include questions that call for short written answers; some also pose problems that require longer written responses. Whatever the form of the questions, your students should have experience with it beforehand so that they are familiar with the format of the test and how they are to respond. But rather than preparing students just before they have to take a test, consider preparing students for assessments as an ongoing part of instruction.

Let's look at multiple-choice test items first. Most adolescents are familiar with them. But not many have spent time in class deconstructing the tests as a way to prepare to successfully complete them. Periodically, put an example of a multiple-choice test question on the board or overhead, listing possible answers as they'll appear in the test booklet. Tell students that they should attempt to find the correct answer by themselves, silently. After a couple of minutes, have students talk in pairs about the answer and how they determined it. Next, ask for volunteers to share. Insist that the volunteers give an explanation of how they found the answer. Always ask if anyone thought about it in a different way. Ask the class if there were any answer choices that they could eliminate immediately, and why. This is a useful strategy to teach students.

You might choose a problem for which the math is familiar to students. Or you might choose a problem for which the math is new. Remember that some standardized tests include items that don't reflect what you are supposed to teach. These items appear on norm-referenced tests as a way to influence the curve of students' results. For example, sixth graders taking a test in late fall might encounter a problem that calls for using angle measurement, something they may not have learned as yet:

Estimate the size of the indicated angle:

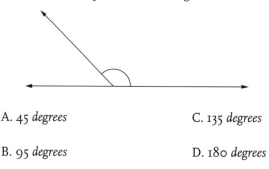

A. 45 *degrees*   C. 135 *degrees*

B. 95 *degrees*   D. 180 *degrees*

Don't think about rearranging your curriculum for items like these. However, it is a good idea to discuss with your students what they might do when they encounter something new and strange. The previous example presents an opportunity for students to think about what they know about angles. In this case, even students who haven't formally studied angle measurement can reason to arrive at an answer by realizing that the indicated angle is considerably greater than 90 degrees but less than 180 degrees. The goal for these exercises is to help your students get used to approaching tests with an attitude of reasoning and making sense of the questions. Mix up the practice to include questions of all the types your students will encounter.

Find out if the questions on your school's standardized tests are ordered by difficulty from easy to hard. Many tests are not constructed this way, so it's helpful to talk with students about persistence. Also find out if answering every question is crucial and if guessing is important.

Tests can be stressful for students, more for some than for others. Tests are also stressful for teachers, as we tend to feel that they are measures of our effectiveness. If you're anxious about the test, don't communicate your anxiety to your students. Instead, let your students know that the test is their opportunity to show their stuff. Encourage them to do their best.

114 How do I help my students prepare for standardized assessments?

Love them or hate them, standardized tests are a part of most students' lives, and important decisions that can significantly affect a student's future are often made based on the results of standardized assessments. Thus, preparation is important. There are several things you can do to help students prepare for standardized assessments so they are not surprised by the content of the test or how the test is formulated.

It's best to devote some time to test-preparation activities regularly, beginning at the start of the school year rather than concentrating them in the few weeks before the test. Take a look at standardized-test practice problems and sample tests, which are often readily available to teachers. Thinking about your current instructional unit, select a variety of multiple-choice test items that relate to various problems or lessons in the unit. Note which item matches each lesson. You don't need to find an item for each lesson, but select enough so that you are able to present one or two items to students per week.

Write each multiple-choice item on a separate overhead transparency. Then, at the end of a problem or lesson, put the appropriate transparency on the overhead and explain to students that this is an example of the kind of questions students are likely to see on standardized tests. The following steps are suggested for helping students make sense of the items:

1. Ask students to attempt to solve the problem alone.

2. Ask students to share their thinking with a partner. Doing so allows them to check out their thinking and increases the likelihood of greater participation in a subsequent class discussion.

3. With the whole class, have students discuss the problem, the correct answer, and their solution methods. Were there answers they could eliminate right away? Try to avoid "teaching by telling" whenever possible, to encourage maximum involvement by your students.

4. Highlight answer distracters (incorrect answers that a student might easily select as the correct answer) and elicit why, mathematically, the test makers might include them.

5. Be sure to discuss how what students did in the lesson they just finished relates to and prepares them to answer the question.

115 What kinds of strategies can I teach my students to use when taking multiple-choice tests?

Here are a few general strategies that students can use to solve multiple-choice test items:

- Decide what the question is really asking.
- Use substitution to find the correct answer.
- Know how to deal with the word *not*.

- Identify and eliminate incorrect answers.
- Use a calculator or other admissible manipulative.
- Draw a diagram, sketch, or table.
- Estimate.

For each strategy, an example problem and sample dialogue is included to help you think about how to support student's learning.

Decide what the question is really asking.

Help students understand what a question means by asking them a series of questions. Consider the following problem:

If a shirt is on sale for 40 percent off, how much will Rosa pay if the regular price is $60?

A. $36   B. $30   C. $56   D. $44

Ask students:

- *Is the question asking what 40 percent of sixty dollars is?* (No.)
- *What is the question really asking?* (How much Rosa has to pay after the discount is figured.)
- *Estimate 40 percent of sixty dollars.* (It should be somewhat less than half of $60.)
- *What answer choices could be eliminated if you understood the question but were unsure how to figure the answer? Why?* (B. $30, because that's 50% of $60 and the question doesn't ask anything about 50%; C. $56, because $56 is just subtracting $4 off of the price. It's not reasonable.)
- *Once you eliminate choices B and C, how could you figure the answer?* (Answers will vary.)
- *How can you check that your answer is correct?* (Answers will vary.)

Use substitution to find the correct answer.

Sometimes substitution is an efficient and effective strategy for solving multiple-choice problems.

Solve for n. $6n + 13 = 43$

A. 2   B. 0   C. 5   D. 14

Ask students:

- *What does it mean mathematically when a number is next to a letter?* (Multiply the value of the letter times the number.)

- *What happens if you substitute answer choice A, or 2, for* n *in the equation?* (6 × 2 = 12, 12 + 13 = 25, not 43, so A is incorrect.)
- *What about substituting answer choice B, zero?* (6 × 0 = 0, 0 + 13 = 13, not 43. So B is also incorrect.)
- *How about 5?* (6 × 5 = 30, 30 + 13 = 43. It works.)
- *And 14?* (6 × 14 = 84. That's already too large.)
- *Why should you substitute every answer choice into the equation, even if it looks like you've found the right answer in choice A or B?* (You might miscalculate.)
- *How would you know that your answer is correct?* (Solve the problem for *n* to check your work. Subtract 13 from each side of the equation and divide the result by 6 to see if you get 5.)

Know how to deal with the word *not.*

Many students get tripped up when a question has the word *not* in it.

Which figure is NOT a parallelogram?

A. ⬭   B. ▭   C. ◇   D. ⬠

Ask students:

- *What might make this question difficult?* (You might not know what a parallelogram is.)
- *What mistakes do you think students might make on this problem?* (Even if you know what a parallelogram is, you may not recognize that the shape labeled C is a square, because it is tilted and therefore a kind of parallelogram; you might think that the question asks which shape is a parallelogram.)
- *Do you have to know that choice D is a pentagon in order to answer the question correctly?* (No. You don't need to know the name of the shape to answer correctly.)
- *Even if you didn't know what a parallelogram was, how could you narrow down the answer choices to the correct answer?* (Look to see what three out of the four shapes have in common, or find a shape that is different from the other three.)
- *What do choices A, B, and C have in common that choice D does not?* (A, B, and C all have four sides. D is the only shape without four sides.)
- *How would you know that your answer is correct?* (D has to be correct because it has five sides and a five-sided polygon is a pentagon, not a parallelogram.)

Identify and eliminate incorrect answers.

It's helpful to know when to use the strategy of eliminating obviously incorrect answers.

The best of these estimates for the height of a door to a classroom is

A. 15 centimeters   B. 2 meters   C. 0.1 meter   D. 2 kilometers

Ask students:

- *What is the question asking?* (About the height of door.)
- *Read the answer choices. Do you notice any that you can eliminate? Why?* (A. 15 centimeters, because a centimeter is about as wide as your fingernail and 15 fingernails wide is clearly wrong.)
- *Any others?* (D. 2 kilometers, because in the metric system, kilometers are used to measure long distances, like how far you drive in a day or how far you run in an hour.)
- *So you've eliminated two answers, leaving B. 2 meters and C. 0.1 meter as your choices. How could you figure out the answer?* (One meter is the same length as a meter stick. If you imagine two meter sticks laid end to end you'd know that it's about right.)
- *Why can you eliminate answer C. 0.1 meter?* (One-tenth of a meter is ten centimeters. If we already know that fifteen centimeters is too short, then ten centimeters is too short also.)
- *How could you check that your answer is correct?* (You could use manipulatives like meter sticks; you could get a meter stick and actually measure the classroom door to help you estimate.)

Use a calculator or other admissible manipulative.

Some standardized assessments allow students to use calculators and other tools. Knowing how and when to use a calculator and when to get a manipulative to solve problems is something every student needs to know.

Which decimal is equal to $\frac{3}{5}$?

A. 0.3   B. 0.35   C. 0.06   D. 0.6

Ask:

- *What might you do if you saw this question on a test and didn't know the answer?* (Use your calculator to help you.)
- *What buttons would you push and why?* ($3 \div 5 =$, because to convert a fraction to a decimal you divide the numerator by the denominator.)

- *How could you check that your answer is correct?* (Answers will vary.)
- *What would you do if you did not have a calculator?* (Divide 3 by 5 using pencil and paper or mentally.)

Draw a diagram, sketch, or table.

Students need to determine when drawing a diagram, sketch, or table is useful.

Billie is sewing fringe around the perimeter of a scarf. The scarf is cut in the shape of an equilateral triangle. One side of the scarf is 9.5 inches long. How much fringe will Billie need?

A. 19 inches   B. 28.5 inches   C. 1 yard, 2 inches   D. 36 inches

Ask:

- *What could you draw to help you understand the question?* (An equilateral triangle.)
- *What is an equilateral triangle?* (A triangle with equal sides and angles.)
- *If you know that the length of each side is equal, how would you find how much fringe is needed?* (Add 9.5 + 9.5 + 9.5 or multiply 9.5 × 3.)
- *How could you check that your answer is correct?* (Divide the other answer choices by 3 to see if they equal 9.5.)

Estimate.

Estimation can be extremely useful on standardized tests. Students need opportunities to determine when they should estimate the answer and when an exact calculation is necessary.

Which is the best estimate for $5\frac{3}{7} + 6\frac{5}{8}$?

A. About 11   B. About 12   C. About 15   D. About 31

Ask:

- *Read the question and the answer choices. Should you calculate the exact answer?* (No.)
- *Why not?* (You're not asked for an exact answer and besides, estimating will help you know about what the answer should be.)
- *How might you estimate the answer?* (You're adding about five and a half plus about six and a half. That makes 5 plus 6, which is 11, and a half plus a half is 1, and 11 plus 1 is 12.)
- *Why might someone be confused and think the answer is A. About 11?* (If they just add 5 + 6 and ignore the fractions, they'd get 11.)
- *Why do you suppose 31 was included as an answer choice?* (If someone made a mistake and multiplied instead of added, they'd get about 31.)
- *How might you check that your answer is correct?* (Answers will vary.)

Ten

Connecting with Parents

The annual fall Back-to-School Night was scheduled, and Greg was considering how to effectively present his math program to his students' parents. "I want parents to experience how their children are learning math and to think about ways to support that learning at home," he thought. "How can I encourage as many of my students' parents as possible to attend? What should I do to assure them that my middle school math class is not an impersonal place? What would be most effective in helping parents understand my math program? How can I let parents know that their students are learning basic skills as well as concepts? What handouts should I prepare?"

Greg considered various activities that would involve parents and give them firsthand experience with the way their children were learning math. He knew that he might not have enough time to involve parents in a math activity, and his thoughts shifted to the many ways that he could connect with parents throughout the year.

Teachers must work to establish credibility with parents. Thus, careful planning for connecting with parents is essential, whether it's at Back-to-School Night, parent conferences, or through other parent-teacher interactions. The more advance thinking and planning you do, the better prepared you'll be, increasing the likelihood of positive interactions with your students' parents.

116 What are some effective ways to connect with parents of adolescents?

There are many ways to connect with parents, including newsletters, Back-to-School Nights and open houses, parent-teacher conferences, homework assignments *(see pages 165–66 for information on parents and homework)*, regular progress reports *(see pages 173–74 for information about assessment and evaluation)*, when parents volunteer in the classroom, and at school "math nights" *(see Question 132 about math nights)*.

Parents of adolescents sometimes indicate that they don't understand what's going on in their child's math class as well as they did when their child was in elementary school. Regular communication is important when students are in middle grades, especially since many adolescents communicate less frequently and in less detail with parents about school-related subjects than they did in elementary school.

117 A math colleague of mine who teaches at another school has no trouble getting most of her parents to attend Back-to-School Night. At my school, however, in a typical year only about a fourth of the parents show up. How can I encourage more parents to attend Back-to-School Night to learn about their child's math education?

It's helpful to examine the factors that might explain low turnout at Back-to-School Night for clues about actions you can take to encourage as many parents as possible to attend. How parents are notified about Back-to-School Night may affect attendance. Relying solely on students to communicate with their parents can contribute to a low turnout if notices aren't getting home. You might choose to mail notices home, put them in the school newsletter, post them on the school Web site, have parent volunteers call parents, and/or have parents sign and return a flier announcing the event. Another factor to consider is that some parents may be intimidated by schools if they have not had positive school experiences. In this case, involving various community organizations such as churches or community centers to get the word out and reassure parents can be helpful. Some parents may not read or speak English. Take advantage of translation and interpretation services in this case. Some parents may not attend because they cannot arrange childcare or because the meeting conflicts with a mealtime. In this case, consider speaking with your principal about providing childcare and refreshments. Finally, if you have any control over when Back-to-School Night is scheduled, check to see that it does not conflict with other events, such as religious holidays or major sports events such as the World Series.

118 What general guidelines will help me prepare for Back-to-School Night?

During Back-to-School Night, many schools organize by having parents follow their child's schedule, rotating classes every twelve to fifteen minutes. Find out how long you'll have with each group and take this into consideration when planning the evening.

Regardless of how your school organizes for Back-to-School Night, the event presents an opportunity for you to show parents how you will impact their child's life throughout the school year. Parents will arrive for the big night with various questions. They want reassurance that your class is a welcoming place for their child, that their child won't be "lost in the shuffle" of middle school, that you care about their child and will provide help when needed. Parents want to know about the curriculum you use, how students are assessed, how much homework will be assigned and how it will be evaluated, and how your class will help prepare their student for future math work. They'll likely want to know about various policies about late work, use of calculators on homework, and so forth. Many are curious about why and how students are expected to work together in partners and groups, to talk with one another about their thinking, and to write about their math learning. Parents also want to know about changes in math standards and instruction. Finally, parents want to know that you are capable of providing students with an effective learning program.

It is important, then, to think ahead about what you want to say to parents, what (if anything) you'd like them to do during that time (such as doing a math activity or viewing and discussing a math video), and what handouts might be appropriate in addressing their questions and concerns. The decisions you make communicate a message to parents. Prepare carefully, to ensure that this message is a positive one!

119 I have enough time at Back-to-School Night to do an activity with parents. What kinds of activities might I include?

Think about what you'd like to communicate with parents. Many teachers want to convey a sense of what learning mathematics in today's middle school classroom is like. The following activities provide a window into what standards-based math looks like.

Sixth-Grade Math

For parents of sixth graders, if you have about fifteen minutes to spend doing an activity, pose *The Horse Problem*. Put a transparency of the problem on the overhead and distribute copies of the problem, along with small slips of paper, to parents:

Dealing in Horses

A man buys a horse for $50.

He sells it for $60.

The man buys the horse back for $70.

He sells the horse again for $80.

What is the financial outcome of these transactions for the man?

Ask parents to silently think about the problem and to record only their answer on the slip of paper. Have parents talk with one another about their answers and reasoning while you collect the papers and record the range of answers on a transparency or on the board.

Financial Outcome for the Man

Made $10

Made $20

Made $30

Broke Even

Lost $10

Lost $20

Lost $30

Show parents the range of answers submitted. Typically, answers include "made $10," "made $20," "made $30," "broke even," "lost $10," "lost $20," "lost $30." Make the point that what's perplexing about this problem is not the numbers—they are easy to deal with—but what to do to determine the correct answer. Ask for volunteers to explain their reasoning for each of the possibilities, beginning with the "lost money" category. Tell participants that you want them to listen to each explanation and see how willing they are to change their opinions in the face of new information, or if they are even more firmly convinced. Acknowledge that sometimes it is hard to listen to other viewpoints when you are invested in a different position, and that now you'd like them to explain why you think a particular answer and line of reasoning is correct. Point out that merely telling the right answer does not necessarily help a student understand why it is correct. You could explain it to them, but that is what they've been doing with each other and it hasn't proved convincing for everyone. Tell parents that another way to look at the problem is to act it out.

Have two hundred dollars in paper money prepared (green construction paper cut up and marked to represent ten-dollar bills works well) in advance. Ask what is needed in order to act the problem out (a horse, a person to buy the horse), and recruit volunteers for each role. Make sure that parents are clear about who has the horse to begin with (not the man!). Give each volunteer one hundred dollars in paper money. To alleviate misunderstandings, you could say the following:

- To all: "*After we act this out, what will we want to know?*" (How much money the man has); "*Who will have the horse at the end?*" (Not the man!)

- To the volunteers: "*Your job is to act out the problem so the rest of us can clearly see and hear what you are doing.*"
- To the audience: "*Your job is to monitor the action to make sure that what they are doing is consistent with the problem.*"

When the action is over, have the "man" count out his money. Explain that in your class, a variety of methods are used to determine answers, and the emphasis is on mathematically sound thinking. Even though parents have heard a variety of explanations and seen the problem acted out, some still may not be convinced about the answer ("made $20"). Let parents know that they still may need time to think about the problem.

Seventh-Grade Math

For parents of seventh graders, you might spend about fifteen minutes on an activity using manipulative materials, such as *Angles with Pattern Blocks and Hinged Mirrors,* to demonstrate to parents how abstract math ideas can be made concrete for students. Working with a partner, parents use hinged mirrors to determine the angle measures of each of the pattern blocks. You'll need one set of hinged mirrors (two small rectangular mirrors taped together so that the reflective sides face each other) and a small handful of assorted pattern blocks for each pair of parents.

Using the information that 360 degrees makes a complete circle, model how the hinged mirrors can be used to measure angles by tucking a square in the corner of a hinged mirror.

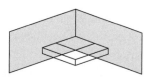

Ask parents to tuck a square in the corner of two hinged mirrors and build the four squares they see using pattern block squares. Ask them to explain why one square corner is one-fourth of 360 degrees, or 90 degrees. Tell parents that they'll be using this method to determine the size of the angles of each of the pattern blocks. Point out that some blocks have congruent angles and others have angles of different sizes. You may want to acknowledge that determining the measure of one angle in particular may appear to be impossible, but that it is possible to figure out the measure of all of the angles. After giving them sufficient time to figure out and record the size of each angle, invite parents to share with the whole group the work they did with their partners. Discuss what students might learn using pattern blocks and hinged mirrors to measure angles of different sizes. Connect this learning to local or state curriculum documents to show parents how this activity fits into an effective instructional program. Some teachers prepare an overhead transparency of a standardized test question that relates to the angle-measuring activity, and talk with parents about how they might use such a question at the end of instruction to communicate with students about how the activity will help prepare them for testing. This provides a natural transition to discussing with parents assessment and evaluation.

Eight-Grade Math

Algebraic thinking is a part of most eighth-grade benchmarks and standards. If you have about twenty minutes with parents, *Toothpick Building* is an activity that can demonstrate to them how children are learning to recognize, describe, extend, and generalize a pattern. You'll need about fifteen toothpicks for each pair of parents. Explain to them that by studying patterns such as the following, students learn about the various uses of variables and how to solve equations. Pose the following question: *If you continue the pattern shown to build a row of one hundred triangles, how many toothpicks will you need?* Explain that the pattern continues in a straight line, not in a loop.

If you continue the pattern shown to build a row of one hundred triangles, how many toothpicks will you need?

Explain to parents that to make the first triangle in the row, you need three toothpicks, and for two triangles, you need five toothpicks. Ask them to make the following chart (you might duplicate the chart for parents) to help them think about how many toothpicks are needed for three triangles . . . four . . . five . . . and so on. Note that organizing data in a table is helpful in looking for patterns.

Triangles	Toothpicks
1	3
2	5
3	

Ask parents to describe the patterns they see verbally. This can help parents connect the table to the equation that describes the relationship between triangles and toothpicks. Give parents time to solve the problem and encourage them to find an equation that would help them determine how many toothpicks were needed no matter how many triangles in a row were constructed, say, 317 triangles in a row, 1,421 triangles in a row, 3,624,501 triangles in a row, or any number of triangles in a row. After sufficient time, ask if anyone was able to find an equation that represented this pattern, and record it on the board or overhead. Parents might volunteer the following:

$y = 2x + 1 \qquad x = \#$ of triangles $\quad$ and $\quad y = \#$ of toothpicks needed

$n + n + 1 = T \quad$ where $\quad n = \#$ of triangles

$T = $ total # of toothpicks needed

Encourage the person to explain how they figured it out. Ask for others to share their equations and methods of finding them. Finally, explain that a coordinate graph makes it possible to represent equations graphically. Put an overhead transparency of a coordinate graph on the overhead or have one prepared on the board. Tell parents that graphing an equation calls for using the ordered pairs to place points on a coordinate grid. Explain that the accepted way to do this is to begin where the two axes meet, called the origin. Remind parents that we want to use the first number in each pair of numbers from the T-chart to count over from the origin and the second number to count up, and place a dot. Plot several points and ask parents to notice that the dots for the triangles and toothpicks are in a straight line, indicating what mathematicians call a linear function. Point to one of the dots and note that the point indicates that _____ many triangles require _____ toothpicks to build it. Ask parents to explain why.

Connect this activity to local or state curriculum documents so parents see how it is part of an effective instructional program. Some teachers make an overhead transparency of a standardized test question that relates to the toothpick activity, and talk with parents about how they might use such a question at the end of instruction to communicate with students about how the activity will help prepare them for testing. Again, this provides a natural transition to talking about assessment and evaluation.

120   I don't have enough time at Back-to-School Night to do an activity with parents. What should I be sure to address?

Parents need to know what their students will learn in math class this year and how they'll be assessed. They also will want to know about your classwork and homework expectations and how students are graded in your class. Finally, parents need to know how to communicate with you about what's going on in math class and how their child is doing in your class. Preparing handouts for parents to take home with them for future reference is helpful.

When I have only a few minutes to spend with the parents in each of my students' classes, I let parents know what I'll be sharing with them in the short time that we have together. I also request that they not ask me how individual children are doing during the evening. I explain that I'm still learning about their child, and that I'm looking forward to meeting with them in a few weeks during parent-teacher conferences to provide a comprehensive picture of their child's progress. I learned to do this after one Back-to-School Night, during which a parent asked me how his child was progressing and I gave him a perfunctory "fine," since I was in between presentations and pressed for time. When I contacted the parent later to discuss some difficulties that his child was having, the parent was understandably upset, since I had told him at Back-to-School Night that his child was doing fine! On the other hand, I do reassure parents that if I have immediate concerns, they will certainly be hearing from me, and that whenever there's a problem that the student and I cannot handle together, I make sure that they know about it.

121 What kinds of handouts about my math program would be most helpful to prepare for parents at Back-to-School Night?

Many middle school teachers prepare handouts to share with parents at Back-to-School Night. Doing so gives parents something to refer to after the evening is over. Consider preparing handouts far enough in advance to allow time for them to be translated into other languages as needed. Include an overview of the instructional materials. Some schools and districts provide a "parent-friendly" version of the district math framework, standards, or course of study, which can be duplicated for parents' reference, giving them ready access to the expectations for their child's grade level. If such a document is not available, prepare an excerpt from the teacher version for parents. Talk about topics they may not be familiar with or recall from their own math learning.

A handout of guidelines for student success in math class can be very helpful for parents. Include information about how your class is structured, your homework expectations, how you evaluate student work, how students are assessed, what students should do to make up work if they are absent, organizational hints including keeping notes and a glossary, and how students can get help when needed. Again, have the documents translated into parents' first languages as needed.

Some teachers have their students write their names on the handouts ahead of time for distribution at Back-to-School Night. (This saves you from having to write 150 or more names!) Then you can send the remaining handouts home to whoever did not attend. Also, it's a good idea to have extra handouts available for families who need them. A sample handout is shown on page 172.

122 A math teacher in my school sends home regular letters to parents. I'd like to try this, but it seems like a lot of work, and I'm not sure all of my students would take the letters home. Is the payoff of using parent letters worth the trouble?

Yes. Parents need to know what you are doing in your math class, why you are doing it, and what math instruction in your math classroom looks like. Parent letters are a terrific way to communicate this. If you're concerned about having students distribute them to parents, post them on a Web site so that parents can access them at their convenience, and send home copies to families who do not have computer access (provide translation as needed). Consider using as a model the sample parent letter included in your curriculum materials. Here's an example of one I sent home to parents outlining a unit focusing on area and perimeter:

Dear Parents,

We are about to begin a measurement and geometry unit that focuses on area and perimeter. Students need many experiences to learn about these ideas and understand how they relate to each other.

In this unit, students work with a variety of materials. They measure the areas and perimeters of regular figures, such as squares, rectangles, and circles, as well as approximate the areas and perimeters of irregular figures. They learn that there are various ways to measure the area and perimeter of both regular and irregular shapes.

Throughout the unit, students use mathematical vocabulary, such as *area, perimeter, square units, circumference, radius, diameter,* and more.

In completing the activities and investigations in the unit, students will also estimate; manipulate whole numbers, fractions, and decimals; and use ideas from the topics of geometry, patterns, statistics, and logic.

Independent activities extend students' experiences in whole class instruction, and regular assessments provide a way to track students' learning. Homework assignments will give you firsthand experience with the unit.

Specific district math standards addressed in this unit include:

- Determine the lengths and areas of simple polygons on graphs.
- Solve problems with whole numbers.
- Solve problems using fractions, decimals, and percents with and without calculators.
- Use formulas for determining the area and perimeter of quadrilaterals (four-sided shapes), triangles, and circles.
- Relate changes in linear measurement to changes in area (for example, when you double the length of the side of a square, the area quadruples).

If you have any questions, please do not hesitate to contact me.

Sincerely,
Cheryl Rectanus

123 How should I prepare for parent-teacher conferences?

During parent-teacher conferences, parents will expect you to provide them with information about their child's progress and grades. While grades are important, a powerful way to demonstrate a student's learning is by sharing samples of student work. These not only help parents understand how their child is performing in your math class, but it gives parents information about what math instruction is like in your class. Select just a few samples of student work that illustrate specific points—you don't need multiple examples and likely won't have time to share a plethora of papers. If you have only a few minutes with each parent, you'll want to prioritize the papers you share so that the most important information is shared first.

Before each conference, look over each sample of student work and think about the points you want to make. You might write a few thoughts on a notepad that you can refer to during the meeting. The notes should include comments about academics and also about the student's behavior and participation in class. Planning ahead will support your conversation with parents during the conference as you focus on their child's progress, strengths, and needs.

Other materials to have ready for conferences include handouts from Back-to-School Night, which provide valuable information about the year's instruction. Be sure you have copies for those parents who need them. Additionally, some teachers use anonymous samples of student work that meet or exceed district standards for their grade level to help parents understand the level of rigor expected.

124 What are effective ways to begin a parent conference? Should I ask parents what questions they have, or just begin discussing how their child is doing?

Begin the conference on a positive note. It's extremely important to convey to parents that you know and respect their child. Then consider asking parents how they think their child is doing in math. Do they see their child sailing through homework assignments, struggling with them, or not doing them at all? Does their child ask questions that reflect an interest in math? This will help give you a broader picture of the student and also give you information about parents' concerns.

Be prepared to respond to the question most typically on parents' minds: "How is my child doing in math class?" This is where you'll share with parents the papers you've selected ahead of time. Explain one work sample at a time, highlighting the points you noted earlier as you prepared for the conference. Offer an anecdote about the student's participation in your class. It might be an insightful comment that the student made during an investigation. Or you might describe an occasion when the student persuaded other students that his or her reasoning was correct. Regardless of the example you choose, parents appreciate hearing something that sets their adolescent apart from others, and they value the individual attention paid by teachers in getting to know their child.

Parents may ask you some questions that you can't answer off the top of your head. This is okay! Don't try to invent an answer on the spot. Tell parents that you need some additional time to think about a response. Be sure that you record their name and question; trying to remember what you promised to whom can be difficult if you are meeting with scores of parents. Then do the research required and get back to them as soon as possible.

125 Some of my students' parents have told me that they dislike math. How can I ensure that they don't pass their attitude on to their children?

It's possible that one or more of your students' parents will approach you during Back-to-School Night and say something like, "You know, I always did poorly in math and hated it. It seems like Debbie is just like me." You probably won't be able to change these parents' attitudes, but you can and should help them avoid passing that negative attitude on to their child.

In addressing this situation, be direct with parents. Tell them that one way they can support their child this year in math class is to be positive about math, even if it is something they dislike. Acknowledge that for some parents, math class was not a pleasant experience. At the same time, explain to them that when they express their dislike of math to their child, that child may come to believe that math is something to be endured—or avoided—rather than enjoyed. Let parents know that this negativity can undermine a child's confidence, and that it is difficult for you as a teacher to reverse the negative perception. Empathize with the struggles that many parents have "revisiting" middle school mathematics with their children. Explain that it is natural for parents to sometimes feel concerned about their ability to help their child with math homework. Reassure them that there are many things they can do to support their child's math learning.

Next, encourage parents to engage their child's assistance in daily activities in which they regularly use math. There are many, many opportunities for this—balancing the checkbook, measuring the amount of wallpaper needed for the bathroom, figuring out how much fertilizer is needed for the lawn, determining dollar savings after a discount has been taken, keeping score when playing a game, and on and on.

126 How can I support parents in helping their child with math at home?

When students enter middle school, many parents feel less confident about their ability to help their child with math homework. And students sometimes want less help from parents, even though they still need it. Addressing this situation can be tricky! Share with parents what you've told students to do when they run into difficulty with homework. It helps to have homework strategies printed on a handout, so parents and students can refer to it at home. Such a support alleviates the criticism that is sometimes unfairly aimed at middle school teachers: "There's no help available for my child when they need it!" What follows is a list of strategies that parents can use in helping students with math at home:

1. Ask your child to reread the problem carefully, aloud. (Alternatively, have someone read it aloud to the child.) Determine what your child understands about the problem and where the confusion lies.

2. Ask your child to explain a problem that they did in class today, and suggest that your child look back at yesterday's (or previous) assignments. These questions/problems may offer hints and provide direction.

3. Have your child call a friend from math class and discuss the problem over the phone. (Some teachers even give out their home telephone numbers to students,

explaining that students may feel free to call for help, if they've followed the first three steps and are still confused.)

4. If the first three steps have been followed and the child is still having difficulty, ask them to write a clear question about the problem to share with the teacher the next day. Parents, include a brief note about the situation for the teacher's review.

Remind parents that today's homework assignments are often quite different from what they received in math class while growing up, and that there's no need to feel embarrassed if they have difficulty making sense of it. Parents may remember math homework as memorizing multiplication tables or incessant practice of a skill that was demonstrated in class; today's homework is often a challenging, open-ended inquiry.

Let parents know that while you hope students are able to complete and understand all of their homework, you recognize that today's problems are more difficult and that not all students will return to class the next day with a correct answer or a clear understanding of the concept(s) involved. Explain that you do expect that students spend sufficient time on each problem so that they are able to participate in the detailed discussions that will take place during the next class period.

Parents should expect that every student will, at times, wrestle with problems and questions. For students to think at higher levels, this grappling is necessary. Explain that you hope that students learn the difference between persistence and frustration, and that you are aiming for the former. *(See Question 137 in Chapter 11, "Handling Homework," for more information.)*

Finally, more and more publishers of math curricula post online homework help for students and parents. Consult your instructional materials to see if this is the case.

127 What if parents tell me that they don't understand the math behind their child's classwork or homework?

First, acknowledge, empathize, and respect parents' efforts to support their child when the student runs into difficulty with an assignment. Reassure parents that they are not alone in this concern, as many parents eventually reach the point where they cannot help with homework in all subjects, not just math. If they reach this point, advise them not to let problems with homework create an undue amount of stress at home. Encourage parents to write a note to you explaining the situation. Let them know that you will gladly make sure that their child gets sufficient help and support. *(See Chapter 11 for additional information on homework issues.)*

128 A parent asked me why math instruction today is different from when they were in school. What might I tell them?

The world around us is changing at an ever-increasing pace. As expressed in the *Principles and Standards for School Mathematics*:

> New knowledge, tools, and ways of doing and communicating about mathematics continue to emerge and evolve. . . . The need to understand and be able to use mathematics in everyday life and in the workplace has never been greater and will continue to increase. . . . In this changing world, those who understand and can do mathematics will have significantly enhanced opportunities and options for shaping their futures. (National Council of Teachers of Mathematics 2000c, 4–5)

Parents need to know that the mathematics that was sufficient for us many years ago is insufficient today. Many years ago, "shopkeeper math" served most adults adequately. Today we are daily faced with problems that require us to think, reason, analyze, and evaluate quantitative information that years ago was only available to a few. Adults today use math to respond to questions such as "Should I get a fixed or adjustable rate mortgage?"; "How should I determine the best retirement plan?"; "What level of life insurance is necessary for my family to continue living within our current means should I die prematurely?"; "How do I understand how much our national debt really is?"; "Should I buy the lottery tickets or not?"; "What's my chance of survival if I undergo this highly experimental treatment?" All of these questions require more than a rudimentary understanding of arithmetic. As a result, statistics and probability, once reserved for higher-level mathematics courses, are now a standard part of the middle school curriculum, as is algebraic reasoning.

129 At my school, eighth graders are placed in either eight-grade mathematics or high school algebra. Parents sometimes ask to have their child placed in the algebra class when my assessments indicate they truly aren't ready. How might I handle this sticky situation?

Every school makes decisions about what classes they will offer students, and in the middle grades, providing several alternatives is not uncommon. It's important that schools give careful consideration to the equity issues inherent in tracking students. These issues notwithstanding, I urge you to consider honoring the parents' request to move their child to a higher-level class if both parents and student are willing to make the effort. If you feel strongly that moving the student to a higher-level class is not appropriate, explain your reasoning to the student and the parents. Share what kind of effort and work quality is required for the higher-level class. Show samples of the student's work and discuss why you feel the placement would not be optimal for the student. Explain the benefits of staying in the current class and let parents and the student know what the student can expect to learn as well as how they can grow mathematically. Be ready to share information about how students can receive additional support in math, whether it is being available to the student before or after school, providing information about support math classes offered at the school, or Web sites that offer math help. Offer to revisit the decision at a logical point in the year, such as at the end of the quarter or semester, and then do so.

130 **I routinely ask my students to explain their thinking both verbally and in writing. A parent challenged me about this, asking, "Why do students always have to explain their thinking? Isn't it the answer that's important?" How might I have best responded to her?**

You might have begun by noting that explaining their thinking supports students' learning. Communicating verbally and in writing encourages students to examine their ideas and think about what they have learned. It helps children deepen and extend their understanding. Both doing mathematics and engaging in writing requires gathering, organizing, and clarifying thoughts, determining what you know and do not know, and thinking clearly. Students' written records are useful during parent conferences as you talk with parents about their child's accomplishments, progress, and needs. When students explain their thinking, you have a better opportunity as a teacher to learn what the student understands and is confused about, which helps you make better decisions about your instruction for their child.

131 **A parent told me that she heard that students no longer had to memorize their multiplication facts. She wondered what had happened to the basics. What should I tell her?**

Reassure the parent that, indeed, students in middle school still need to know their multiplication facts! Most parents are referring to arithmetic skills when they mention the basics. Most of what many parents remember is doing pages of worksheets with computation problems, perhaps coupled with some word problems. Today's curriculum, however, includes a broader notion of arithmetic. Talk with parents about what is basic to math instruction today, and reassure them about the importance of arithmetic and how their child will be helped to learn the basics. In addition to being computationally proficient—whereby students add, subtract, multiply, and divide whole numbers and fractions, decimals, and percents efficiently and effectively—students today need to know how to apply computation skills. This includes

1. selecting the operation needed;
2. choosing the numbers to use;
3. performing the calculation, either using mental math, paper and pencil, or a calculator;
4. evaluating the reasonableness of the answer (which requires being able to estimate).

Explain that applying computational skills to solve problems and developing good number sense are essential parts of math learning and that in your classroom, students will have many opportunities to develop and refine basic skills.

132 Some of my colleagues at other schools hold "math nights" at school. What do these events involve, and what are ways of ensuring their success?

Math nights are events that involve the school community in celebrating math and highlighting its importance. They are a wonderful way to convey to families your instructional goals and enthusiasm for math learning. Don't try to organize them alone, however. Find at least one other person at your school to assist with the planning and execution of the event, such as a colleague or an administrator. Some schools present a series of math nights. Whether your school decides to present such events singly or in a series, preplanning is essential.

Purposes

It's important to be clear about the purposes of the event. Are you trying to involve students in planning activities that reinforce their math learning? Increase parent involvement? Show parents what students are learning? Being clear about the purpose(s) will help you set goals for the evening.

Goals

Perhaps you want to address parents' fears about math and help them feel confident in helping their children at home. Maybe you want parents to learn about the math curriculum. Perhaps you'd like to survey parents to learn what types of math-related activities they'd like to participate in. You probably want to encourage a large turnout and involve students in planning activities. Whatever your goals, they should drive your decisions about the kinds of activities selected for the event.

Outreach strategies

Consider ways of announcing the event. Here is a list one school brainstormed in preparation for a math night at their school:

- Make invitations and posters.
- Translate invitations into parents' first languages.
- Share the information with the principal and school staff.
- Announce the event(s) at school alliance meetings (e.g., PTSA, etc.).
- Determine ways to generate interest. (For example, you might put a large jar of candies in the hallway a couple weeks ahead of time and invite passersby to estimate how many candies are in it. Next to it, place a smaller jar of candies, with a sign indicating how many candies are in that jar, as a referent for making an estimate. Stack fliers announcing the math night next to the jar, explaining that the number of candies will be revealed at the math night, and the contents given to the person or people with the closest estimate.)
- Invite parents and students from local elementary schools.
- Announce the event in school newsletters.

Partners

Who might you partner with at the math night? Do you have school or community groups who would like to host a table and who would be willing to help get the word out about the math night through their organization?

Preparing students to assist at the math night

Many teachers involve their students in selecting and leading appropriate activities for the math night. This is a wonderful way to help students take ownership of and think carefully about the mathematics they've been learning.

Details, details, details . . .

Other nuts-and-bolts items to plan for include: providing refreshments; arranging for interpreters; determining door prizes; providing childcare for families with young children; setting up and cleaning up; getting equipment, such as tables, chairs, a microphone, an overhead projector, and so on; creating and duplicating handouts and an evaluation form; applying for necessary permits for use of the building; and so on. Be sure to determine who is responsible for what task, and communicate the expectations clearly.

Building the agenda

Think through the timing of the event and each participant's role. One school brainstormed the following list:

5:45 P.M. – 6:00 P.M.

The principal talks about the importance of math, the importance of supporting students, and shares the new district-wide expectation that all ninth graders will take algebra or a higher-level course beginning the next year.

6:00 P.M.– 6:30 P.M.

A curriculum-related activity gives parents an opportunity to see what math class is like for students.

- Students plan and lead the activities based on criteria set by teachers. Teachers review Middle School Family Math, district curriculum, and other resources for students to draw from.
- Families receive "passports," which are stamped for each activity they participate in. They also receive tickets that they can submit for door prizes.

6:30 P.M.– 7:00 P.M.

During dinner, speakers—teachers and invited guests—focus on how parents can support learners in math.

- Handouts are distributed to parents (on tips for helping your student, summer opportunities for middle school students in math-related-areas, etc.)
- Community partners offering students extended learning opportunities are introduced.

7:00 P.M.– 7:15 P.M.

- Distribute door prizes and thank families for coming.

7:15 P.M.

- Clean up
- Gather data (such as number of families attending) for later analysis.

Grading Policy and Classroom Expectations

Teacher: Cheryl Rectanus, RM. 300   **Phone:** (W) 555-1234 (H) 555-4321 until 9 P.M.
email: crectanus@anyemail.org

Homework Web site: anygrademath@homeworkwebsite.org

About Oregon Math Standards . . .
Our class is organized to prepare students to meet or exceed Oregon State Standards in math. Topics include Algebraic Reasoning, Number and Computation, Geometry, Statistics and Probability, and Measurement. Assignments are accessible for the least experienced learner and provide opportunities for those who need additional challenges.

What math program is used?
"Terrific Math Program." It's a highly-rated, classroom-tested middle school program aligned with district, state, and national math standards. It addresses a broad range of math topics and emphasizes math reasoning, problem solving, the development of strong number sense and computation, and communication of mathematical ideas.

Is homework assigned?
Yes, three to four times a week and occasionally on weekends, approximately thirty minutes per assignment. Students should have access to a calculator; some assignments will indicate "no calculator."

How do I know what my child's homework assignment is?
Students keep track of assignments and due dates in their school planner.

What if my child gets "stuck" on their homework?
Have your student call a classmate. If help is still necessary, call me at home until 9 P.M.

When are assignments due?
They're due the next school day, unless otherwise specified. Check your child's planner.

Is late work accepted?
Yes, with a note from parents.

How can I know if my child has turned in an assignment?
Ask to see their planner. Assignments with a check mark next to it were completed. Circled assignments were not handed in. Also, I send home progress reports every two weeks.

(continued from page 172)

How are students assessed?

Several types of assessments are used to determine what students know and can do:

- *Classwork and homework.* These assignments provide me with helpful information about how your child approaches and solves math problems.
- *Work Samples.* Any assignment scored using the Oregon State Scoring Guide.
- *Quizzes* and *Unit Tests.* Sometimes these are short-answer, other times more lengthy responses are needed. Students demonstrate their understanding of concepts, skills, and vocabulary.
- *Formal assessments given to all Oregon eighth graders.* These include the Oregon Math Knowledge and Skills Test (MKST) and the Math Problem Solving Assessment (MPSA). The MKST is a multiple-choice test covering all five math topics. A score of 231 meets the standard. The MPSA requires students to choose and solve one of two problems. The student demonstrates, as clearly as possible, his or her understanding of the problem, the method(s) used to solve it, why the method makes sense given the task, and how they verified their method and answer. The MPSA is scored using the Oregon State Scoring Guide; your child has a copy of this in his or her math notebook. To meet standard, the student must earn a minimum score of 32.

How are students graded?

Students are graded on their achievement and their effort. Students demonstrate their achievement through classwork, homework, quizzes, tests, and work samples. Effort is demonstrated in participation in individual and group assignments, class participation, assignment completion, and willingness to revise work. Students are encouraged to revise work and to turn it in again for reevaluation. Classwork, homework, quizzes, and tests are scored using the following performance levels:

E = Exceeds (Student demonstrates and applies strong academic performance that is above grade-level standards.)

M = Meets (Student demonstrates academic performance that meets grade-level standards.)

C = Close to meeting (Student demonstrates academic performance that is close to meeting grade-level standards.)

N = Does not yet meet (Student demonstrates academic performance that is below grade-level standards.)

(continued from page 173)

NE = No evidence (student did not submit enough work to determine a mark for academic performance.)

Effort levels include:

A = Almost Always (student always demonstrates an effort to learn, consistently completes work, and is diligent and on task)

O = Often (student often demonstrates an effort to learn, usually completes work, and is usually diligent and on task)

T = Sometimes (student sometimes demonstrates an effort to learn, inconsistently completes work, and is inconsistently diligent and on task)

S = Seldom (student seldom demonstrates an effort to learn, infrequently completes work, and is seldom diligent and on task)

Are students required to learn the basics?

Yes. Number sense and efficient computation strategies are important in today's world!

How can I help my student stay organized?

Students are required to keep a math section in their binder that includes their math book, information about math class, a glossary, and notes from classwork and homework. Ask to see your child's binder. If it looks empty or disorganized as the year progresses, you should be concerned. Drop by anytime before or after school to see a sample binder for student and family reference.

Eleven | *Handling Homework*

Problems, projects, posters, packets, worksheets, interviews, games," Nancy remarked to her colleague Andy. "So many ways to assign homework! Parents and students expect homework and my school requires it. But what makes most sense for my students? And how do I manage the paperwork generated by almost two hundred kids and still have a personal life?"

"I know what you mean," said Andy. "There never seems like there's enough time! I'm wondering how I'm ever going to get through everything I need to teach this year when I only have forty-two-minute class periods."

"In that case, why don't you use homework as a way of extending the lesson?" Nancy remarked. "You could prepare ways for your students to extend certain lessons for homework. Then you can use their homework to launch the next class period."

These comments reflect how important it is for teachers to think about how learning goes on outside the classroom. Why do you assign homework? What sorts of homework assignments can support learning? What should you do with the homework assignments your students complete? And what can you do about students who don't do their homework? It's important to think about the role of homework in your math instruction and how to communicate about homework to students and parents. It's also important to remember that not all students have access to homework help outside of school. What kinds of organized support might your school provide through after-school homework help clubs, and parent or older student tutoring or study sessions?

133 What are the purposes for assigning math homework?

Most of us remember doing math homework when we were students. For many of us, homework consisted of a daily diet of worksheets and drills. I can remember once in eighth grade asking my teacher why we had to do so much math homework; she replied "simply because parents expect it." I remember resenting the fact that she was just trying to impress parents, as if the quantity of homework would be an indication to them that her math class was rigorous and challenging. While most families do expect homework to be assigned in math class, it's important for you to carefully consider the purposes for assigning homework and the potential benefits for student learning.

Some teachers use homework assignments to help students reinforce something that they learned in class. Other teachers prefer to assign homework that requires students to apply a concept or skill in a new situation. Sometimes homework is used for students to extend their thinking about something learned in class. Some teachers ask their students to collect data for homework that will be used in class the next day. It's important to keep in mind that homework is also a useful vehicle by which students can communicate with their parents about what they are learning in school. *(See Chapter 10 for more information about parents and homework.)*

There's no single "best" or "right" way to think about homework. Consider the purpose you want it to serve for students. This will help you decide what assignments make the most sense. Do you want students to apply what they've learned in a new situation? Are you trying to help students connect their learning to an earlier concept, activity, or unit? Is the intention skill maintenance? Or is the goal to extend something that students learned in class? Different purposes call for different types of assignments. Be sure that you've considered the level of complexity of a homework assignment. If you're assigning a problem to be solved, for example, it's important that you've tried it yourself. If the assignment calls for gathering data, make sure the activity is reasonable. You don't want homework to create undue problems for your students or their parents.

Don't use homework as a punishment for misbehavior in class. You don't want students to associate something that should be positive and rewarding with misbehavior. And don't excuse students from homework as a reward for productive behavior. It sends the message that homework is drudgery and should be avoided.

134 Is it beneficial to always assign the same kind of homework, or should I introduce more variety?

Some teachers select potential homework assignments in the context of planning a unit of instruction. By having a repertoire of appropriate possibilities upon which to draw from, you can pick and choose assignments based on what makes most sense for students. This approach also allows you to assign homework on the spot, when the lesson presents something that you'd like your students to pursue. Having your students answer a question or explore a conjecture that arises during a class discussion can make an effective and relevant homework assignment. It's good to be adaptable with respect to homework. For example, when planning a unit on computing with fractions, Rose selected several homework assignments that would be appropriate for the unit. During a discussion about multiplying fractions one day, however, a student commented, "When you multiply two whole numbers the product always gets bigger. But it seems like when you multiply two fractions, the product always gets smaller. Is that always true?" Rose, who had already prepared a homework assignment for the class, saw the opportunity for learning that had been presented and changed her plans, asking the class to investigate the student's question for homework instead.

Other teachers prefer to routinize homework. Cheryl Ann, for example, assigns homework twice a week. Her Monday assignment is due on Wednesday. On Tuesday, she asks the class how they're doing with the assignment, checking to see if the vocabulary or language is creating a problem for students and to see if clarification is necessary. Cheryl Ann asks the class what challenges they're having with the assignment. Usually, students have questions. If they do not, she asks the class to take out the assignment and reread it. Cheryl Ann gives a new assignment on Wednesday, checks with students about it on Thursday and has students turn it in on Friday. The advantages to this system are that students and parents know exactly when homework will be assigned and how the routine will be handled each day. A disadvantage is that the teacher has less flexibility to assign homework based on something that happened in class that day that she'd like the class to pursue. Also, some students find such a routine to be monotonous.

Melissa assigns her students a "problem of the week." In this longer-term homework assignment, she selects a more complex problem that requires more time than overnight to complete. Each day, Melissa checks with the class about how they're doing on the assignment. She asks students to report what they're finding out about the problem so far. At the end of the week, students report their findings and summarize the problem. A long-term problem such as a "problem of the week" allows students to deeply investigate a problem or idea.

135 How long should it take students to complete their math homework?

Go over the problems you are thinking of assigning to get a sense of how long it may take students to complete them. Some teachers use what's commonly known as the "thirty-minute rule." This means that you assign work that you expect students to complete within thirty minutes, give or take a few minutes. You want students to make an honest attempt on the assignment, but don't want them spending hours on it. You also don't want students to report tears or frustration from the volume of work you've assigned. When you are starting out, you have to make educated guesses about how much homework to assign, letting students' reactions and the quality of their work inform those decisions. You'll find yourself (and your students) getting better at dealing with homework over time.

136 Getting students to complete their math homework is difficult. What are ways to encourage students to complete assignments?

It's important to think through assignments ahead of time, so you know that they are worthy of students' time. Be direct with the class about the purpose of the assignment. Students need to feel that you're asking them to spend their time on something that's of value. They also need to see the connection the assignment has to their classroom learning. If adolescents perceive the assignment as busywork, they are less motivated—and less likely—to complete it.

Make sure that your directions are clear. More students complete homework when they understand what's expected of them. It helps to either duplicate the assignment for students or have them record it on an assignment pad, in their binder, or some other place that is not likely to be misplaced. Rather than simply writing on the board something like *HW: pp. 122–23 #1–4*, and then dismissing class, spend a few minutes before the end of class discussing the homework assignment and going over your expectations. Ask a volunteer to read the assignment aloud to the class, and follow this by having several students tell you in their own words what they think the assignment is. Prompt students: What materials will they need? How might they get started? What can they do if they get stuck? These are helpful questions to ask. Some teachers ask students to check in with a neighbor to clarify the assignment. Finally, give students an opportunity to ask questions about the assignment.

For example, Kacy's sixth graders were learning about area and perimeter. For homework, they were asked to draw three shapes, each with a perimeter of thirty centimeters, which they'd use in an activity the next day. Students were to draw their shapes on centimeter-squared graph paper following two guidelines: draw only on the lines, and draw shapes that can be cut out in one piece. Kacy introduced the homework assignment by passing out a sheet with the directions printed on it as well as a sheet of centimeter-squared graph paper.

Homework

The Perimeter Stays the Same

You need: centimeter-squared graph paper

To do:

Draw three shapes on your centimeter-squared graph paper. Each shape must have a perimeter of 30 centimeters and follow these rules:

- Draw only on the lines.
- Draw shapes that will remain in one piece if you were to cut it out of the paper.

Due: tomorrow

She asked a student to read the directions aloud. When the student finished, Kacy asked the class to explain in their own words what they were expected to do. She spent some time eliciting from students what "perimeter" is, and how it is different from "area." Kacy then placed a transparency of a sheet of centimeter-squared graph paper on the overhead.

"May I have a volunteer to come to the overhead and draw a shape that follows the rules?" she asked.

A number of hands shot up. Kacy called on Lynn, who came to the overhead and drew the following shape:

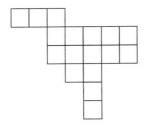

"What are the rules the shape has to follow?" Kacy asked.

"It has to stay on the lines," several students volunteered.

"Does it do that?" she asked.

"Yes!" the class chorused.

"And the other rule is the shape has to stay in one piece if you cut it out," volunteered another student.

"Would the shape that Lynn drew follow that rule?"

Another resounding "yes" was heard.

"What about the perimeter? Is the perimeter thirty centimeters?" Kacy asked.

"No, it's only twenty-six centimeters," replied Lynn after she'd counted the perimeter intently as the class watched. "Let me add a bit more." Lynn continued adding squares to the drawing until it looked like this:

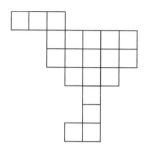

Lynn counted carefully and confirmed that the perimeter was thirty units.

"So it fits the rules. What questions do you have about the homework assignment?" Kacy asked the class. There were no questions, and students were dismissed.

"I can do it now," Kacy overheard a boy say as he left the classroom. "It really helped to see it drawn on the overhead."

Giving students choices can also be helpful. For example, you might explain to the class, "For homework there are seven questions. Everyone is to answer numbers one and two. Then choose three more of the remaining questions to answer." Of course, questions 1 and 2 should be questions you really want everyone to do. Students indicate that they like having choice about their assignments and control over their learning. For many, this is highly motivating. Again, spend some time introducing students to the problems so they understand what they are about and what is expected.

Finally, check with your students' other teachers about their homework assignments. If you know that a major science project and social studies report are due, ease off on your assignments that night and ask for the same courtesy from them when you know you'll be assigning an especially complex or long assignment.

137 What should I tell students to do if they get stuck while they're working on their homework? How can I involve parents in helping them?

Students often wonder what they should do when they are stuck. And many parents say that while helping their child with math homework was easy in elementary school, now that the math their student is doing is more complex, they aren't as confident as they once were in their ability to help. *(See Chapter 10 for more information on addressing parental concerns with respect to homework.)*

In my classroom, I've shared the following guidelines with students and their parents:

1. *Figure out what you understand and what you don't understand.* It's easier for someone to help if the helper knows what you are confused about.

2. *Check to see if there is something in your text, or a previous assignment, that can help.* This is a good reason to keep your work organized in your binder, so you can look back at it when you need it.

3. *Ask an adult, if possible, for help.*

4. *Call several classmates.* I set a minimum of three.

5. *Check our classroom homework Web site.* Some teachers post assignments online with suggestions about what to do if the student runs into difficulty.

6. *Call me at home.* Students and parents are often relieved to know that they can reach me directly if they really need help. Setting limits is, of course, important. I explain to students that they can call me up until 9:00 PM, and they should be prepared to tell me which of their three classmates they spoke with about the assignment before calling me. Even with class loads of 160 or more students, I don't typically receive more than a few calls on any given assignment.

7. *Come in before or after school for help.* Some teachers hold a regular "Homework Club," where students can meet with them and other students to work on the assignment together.

138 My class periods are short, and reviewing homework that students have completed sometimes takes too much class time. But simply collecting and grading the assignments doesn't seem to motivate my students to complete their homework. What are some effective ways to deal with completed homework?

There are several ways to handle completed homework. What you do with it depends on why you assigned it. Processing the assignment in some way, however, is important.

If an assignment was designed to give students practice with something they've already learned, you might ask students to share their answers and strategies for solving their homework problem with a group of three or four students. Explain that the group is expected to come to consensus about the answers. Keep the focus on understanding the assignment, not merely on getting right or wrong answers. After sufficient time has passed (monitor the students so you know when most groups are done), ask the class what questions they want to bring up for the whole class to discuss. Discuss only those questions.

Sometimes teachers assign homework that will be processed during a later class discussion. For example, say that the class was learning about probability and you had assigned them the problem *Two-Coin Toss* for homework. The problem goes like this:

Homework
Two Coin Toss

You need: two different coins

To do:

1. Toss two coins together 25 times.
2. Keep track of what comes up on each toss.
3. What do you think will occur most often? Note: when you toss two coins, there are three possible outcomes: two heads, two tails, or one of each.

Due: tomorrow

When the class arrives the next day, have a chart ready for students to record their data. Explain to the students that while they are working today, they should take time to record their homework data on the class chart that you've prepared on an overhead transparency:

Two Coin Toss

Use tally marks to record your homework results below.

Two Heads	One of Each	Two Tails
l l l l	l l	llll

Such an assignment is motivating to many students, provides data for a subsequent class discussion, and is a nice alternative to collecting and marking homework papers. Many teachers report that when they assign and simply collect homework to grade and hand back later, some students are not motivated to complete it. When a student knows, however, that doing their homework directly affects their experience in class the next day, many more will complete it.

139 I'm overloaded with completed homework assignments! Is it really necessary to collect and read every assignment?

With class loads of upward of 160 students, it is simply not practical to collect and score every homework assignment, let alone every classwork assignment. Advance planning is helpful. Look at the assignments you've planned for a given unit or term. Which are likely to convey to you where students are with their learning? You will want to look at these carefully for each student. Which are designed primarily to provide opportunities for students to practice or review what they've learned? These can be dealt with in class. Decide which assignments fall into each category. Consider alternating which class you collect work from so you're only taking home one or two sets of work rather than six.

140 What should I do if students don't do their homework?

What to do about students who don't complete their homework assignment depends on the kind of assignment given, how you plan to follow up, and how many students didn't do the work. The frequency of the behavior must also be considered.

If only a few students neglected to do their homework, and it happens infrequently, it's probably a minor issue that you can address easily. If students share their homework in pairs or small groups, then it makes sense to have those students who didn't complete the assignment at least talk with others who did. This keeps them involved in the learning, which is the main goal of assigning homework in the first place. For everyone to agree on the results, some students may have to work a bit harder to include someone who didn't do the assignment. But doing so can reinforce their learning as well.

For students who habitually do not complete their homework, other approaches are appropriate. You will want to get in touch with the parents of chronic offenders to discuss the situation. A parent-teacher conference in which the student is also present can be helpful in addressing the reasons the student is not doing their homework, and for suggesting ways to support the student, such as a written agreement between the student, their parents, and their teacher about what the student will do and rewards or sanctions for compliance or noncompliance, an after-school homework club, or a tutor if needed.

If a significant number of students haven't done their homework, it's possible that the assignment you gave wasn't clear or was too difficult, given the knowledge they currently have on the topic. Or it could be that students have not yet internalized your expectations for homework completion. Either way, address the situation immediately with the class. Encourage an honest discussion, so that you can determine what went wrong. If the directions were unclear to your students, a simple clarification may suffice. Perhaps, after some reflection, you will decide that the assignment you gave is more appropriate for in-class work. Another possibility is that the homework assignment wasn't appropriate in the first place. You may want to revise it in some way and give it a try as another assignment. Finally, in some schools, doing homework is not part of the school culture. Addressing this problem requires a systemic approach involving the administration and other teachers in the school.

141 What about late homework? Some students don't do any homework during the quarter and then hand in a blizzard of math papers for me to score before the end of the grading period.

Homework is a way to encourage students to learn beyond the bell. Typically, assignments are relevant to the current unit of study, and many teachers won't accept work turned in more than a week after the due date. Sometimes, though, teachers make exceptions and accept late work beyond that point. However you choose to handle late work, it's important that you clearly communicate your homework policy with students and their parents; make clear your expectations and the consequences for incomplete or missing homework assignments. If and when a student neglects to do or return homework on a regular basis, it may be an indication of a more serious issue that you need to address. If this is the case, you should do so swiftly and tactfully.

142 How should I score the homework that my students complete?

Some assignments warrant scoring for academic achievement; others do not. Keep in mind the purpose of the homework assignment when making this decision. Many teachers score homework for effort or participation rather than for achievement because they aren't really sure who has done the work, the student or the student's parents. Homework intended to summarize what a student has learned from a unit might be appropriate for scoring for academic achievement. Practice or new topics may be more appropriate to score for effort rather than achievement. If you are going to score the papers as evidence of academic achievement, it's appropriate to do so after students have discussed the completed homework in class and had time to revise their thinking based on the discussion. *(For more information on assessment and scoring, see Chapter 9.)*

143 I teach 160 students a day. How do I organize the homework they turn in?

Without an efficient system in place, it is frustrating and overwhelming trying to deal with the homework generated by many students. When you decide to collect a homework assignment, here are a few ideas for organizing the papers:

1. Have one student from every table collect the work and put it in an in-basket or folder designated for that class.
2. Keep a tub labeled with the class period near the door and have students place their work in it on the way out the door.

Decide what you want students to do with their homework assignments. Do you want them to keep them in a binder for future reference? In a portfolio that stays in the classroom? Thinking through these questions will help you decide what to do with students' homework.

144 I like the idea of having students display their homework in class, but I don't want the next period's class to see it. How might I deal with this situation?

Have a routine in place whereby students take down their work and put it in a special place in the classroom at the end of the period. If students are assigned roles for classroom management, it can be the "resource manager's" job to retrieve and repost the work at the beginning of the period and take it down as needed at the end of class.

Some teachers have students use butcher paper to cover over work that is posted. Other teachers don't worry that students will closely examine the work from another class.

145 My eighth graders don't know their multiplication facts. Should I make them memorize them for homework?

If students understand the concept of multiplication and "know" most of their facts, memorizing those they don't know is in order since fluency is important in mathematics. *(See Chapter 2 for more on students learning the basics.)* One teacher handled this situation in the following way. Terry introduced her class to a three-week memorization project. To begin, she wrote the equations (without answers) from 0×0 to 12×12 randomly on a sheet of paper. She photocopied it and distributed copies to each of her students, instructing them to write the answers to as many problems as they could in six or seven minutes, without using a calculator. When the time was up, she had them sign their papers, collected them, and later circled those the students had answered incorrectly or left blank. She photocopied the students' papers for later reference. The next day, Terry handed back the original papers and told her class that during the next three weeks, they would be participating in a project in which they would be memorizing different things. Those who had some multiplication facts to learn would memorize them. Those who knew all of their multiplication facts would memorize other things—such as common measurement equivalents, the first ten prime numbers, common percent-decimal-fraction equivalents, the first ten square or triangular numbers, and so on. *(Note:* it is important that students have direct experience with concepts before memorizing them since memorizing without understanding can negatively affect later learning.) Those in the second category (who knew all their facts) had to tell Terry what they were going to memorize. Next, Terry taught the class how to study and provided them with three or four techniques for committing something to memory. She finally told the class that they would spend five minutes in class studying each day for the next three weeks, and on each Friday, students would have five minutes in which to record every fact that they had agreed to memorize. This paper would be graded. The process would repeat for three consecutive weeks.

Every Friday, Terry had students record from memory what they were to memorize, and she collected and graded the papers according to what the student had agreed to memorize. She told me that she made the project three weeks long so students had enough repetition for the facts to become automatic. She explained that the first year she tried the project, she allotted only one week for it, but a week wasn't long enough for students to remember what they had agreed to learn. Furthermore those students who at the end of the week demonstrated that they needed further practice didn't get the practice because the project was over! So the following year Terry lengthened the project to two weeks, and that still seemed too short. She finally extended the project to three weeks, which seemed about right. Most students managed to memorize the multiplication facts they didn't know yet in that time.

For those who hadn't, Terry took a careful look at which facts they were missing, and tried to determine if the student had a studying issue or if the student didn't understand conceptually what was going on with multiplication. For those students, Terry provided additional support. She gave them a multiplication table and asked them several questions in an interview format. Terry asked the students to tell her about the table and how it worked. She asked them if they knew what happened when you multiply a number by zero, what the resulting product would be. If the student could tell her that it was zero, Terry said, "So you know your zeros multiplication facts. Let's cross them off the table." She proceeded to draw a line through the zero row and column. "What about the ones multiplication facts? What happens when you multiply a number by one?" If the student answered correctly, she had them draw a line through the ones row and column.

X	0	1	2	3	4	5	6	7	8	9	10	11	12
0	0	0	0	0	0	0	0	0	0	0	0	0	0
1	0	1	2	3	4	5	6	7	8	9	10	11	12
2	0	2	4	6	8	10	12	14	16	18	20	22	24
3	0	3	6	9	12	15	18	21	24	27	30	33	36
4	0	4	8	12	16	20	24	28	32	36	40	44	48
5	0	5	10	15	20	25	30	35	40	45	50	55	60
6	0	6	12	18	24	30	36	42	48	54	60	66	72
7	0	7	14	21	28	35	42	49	56	63	70	77	84
8	0	8	16	24	32	40	48	56	64	72	80	88	96
9	0	9	18	27	36	45	54	63	72	81	90	99	108
10	0	10	20	30	40	50	60	70	80	90	100	110	120
11	0	11	22	33	44	55	66	77	88	99	110	121	132
12	0	12	24	36	48	60	72	84	96	108	120	132	144

They would continue with the fives, tens, twos, fours, sixes, threes, doubles, and so on. She asked about the facts in this order because she wanted the students to see that if you knew your twos, you could use that information to figure out the fours, and so on. By the time they had finished crossing out the facts the student knew, there were only a few facts left to memorize.

X	0	1	2	3	4	5	6	7	8	9	10	11	12
0	0	0	0	0	0	0	0	0	0	0	0	0	0
1	0	1	2	3	4	5	6	7	8	9	10	11	12
2	0	2	4	6	8	10	12	14	16	18	20	22	24
3	0	3	6	9	12	15	18	21	24	27	30	33	36
4	0	4	8	12	16	20	24	28	32	36	40	44	48
5	0	5	10	15	20	25	30	35	40	45	50	55	60
6	0	6	12	18	24	30	36	42	48	54	60	66	72
7	0	7	14	21	28	35	42	(49)	(56)	(63)	70	77	(84)
8	0	8	16	24	32	40	48	(56)	(64)	(72)	80	88	(96)
9	0	9	18	27	36	45	54	(63)	(72)	81	90	99	(108)
10	0	10	20	30	40	50	60	70	80	90	100	110	120
11	0	11	22	33	44	55	66	77	88	99	110	121	132
12	0	12	24	36	48	60	72	(84)	(96)	(108)	120	132	144

Bibliography

Allen, D., and T. Blythe. 2004. *The Facilitator's Book of Questions*. New York: Teachers College Press; Oxford, OH: National Staff Development Council.

Arter, J., and J. McTighe. 2001. *Scoring Rubrics in the Classroom*. Thousand Oaks, CA: Corwin Press.

Ball, D. L., and D. K. Cohen. 1999. "Developing Practice, Developing Practitioners: Toward a Practice-Based Theory of Professional Education." In *Teaching as the Learning Profession: Handbook of Policy and Practice*, edited by L. Darling-Hammond and G. Sykes, 3–32. San Francisco, CA: Jossey-Bass.

Becerra, A. M., and J. Wesiglass. 2004. *Take It Up: Leading for Educational Equity*. Santa Barbara, CA: National Coalition for Equity in Education.

Boaler, J., and C. Humphreys. 2005. *Connecting Mathematical Ideas*. Portsmouth, NH: Heinemann.

Bresser, R. 2003. "Helping English-Language Learners Develop Computational Fluency." *Teaching Children Mathematics* 9(6):294–99.

Bresser, R., and C. Holtzman. 1999. *Developing Number Sense, Grades 3–6*. Sausalito, CA: Math Solutions Publications.

Burns, M. 2000. *About Teaching Mathematics: A K–8 Resource*. 2nd ed. Sausalito, CA: Math Solutions Publications.

———. 2004. "Writing in Math." *Educational Leadership* 62(2):30–33.

Burns, M., and C. Humphreys. 1990. *A Collection of Math Lessons from Grades 6 Through 8*. Sausalito, CA: Math Solutions Publications.

Burns, M., and R. Silbey. 2000. *So You Have to Teach Math? Sound Advice for K–6 Teachers*. Sausalito, CA: Math Solutions Publications.

Chapin, S., C. O'Connor, and N. Anderson. 2003. *Classroom Discussions: Using Math Talk to Help Students Learn*. Sausalito, CA: Math Solutions Publications.

Chazen, D., and D. L. Ball. 2001. "Beyond Being Told Not to Tell." *For the Learning of Mathematics* 19(2):2–10.

Cohen, E. 1994. *Designing Groupwork.* 2nd ed. New York: Teachers College Press.

Danielson, C. 1997. *A Collection of Performance Tasks and Rubrics.* Larchmont, NY: Eye on Education.

DeBell, M. 2005. *Rates of Computer and Internet Use by Children in Nursery School and Students in Kindergarten Through Twelfth Grade: 2003.* NCES 2005–111. Washington, DC: U.S. Department of Education National Center for Education Statistics.

Echevarria, J., M. E. Vogt, and D. J. Short. 2004. *Making Content Comprehensible for English Learners: The SIOP Model.* Boston, MA: Pearson Education.

Fay, J., and D. Funk. 1995. *Teaching with Love and Logic.* Golden, CO: Love and Logic Press.

Hiebert, J., T. Carpenter, E. Fennema, K. Fuson, D. Wearne, H. Murray, A. Olivier, and P. Human. 1997. *Making Sense: Teaching and Learning Mathematics with Understanding.* Portsmouth, NH: Heinemann.

Lambert, L. 2003. *Leadership Capacity for Lasting School Improvement.* Alexandria, VA: Association for Supervision and Curriculum Development.

Lane County Mathematics Project. 1983. *Problem Solving in Mathematics: Grade 7.* Menlo Park, CA: Dale Seymour Publications.

Lappan, G., J. Fey, W. Fitzgerald, S. Friel, and E. Phillips. 2000. *Variables and Patterns.* Menlo Park, CA: Dale Seymour Publications.

Lesh, R., T. Post, and M. Behr. 1988. "Proportional Reasoning." In *Number Concepts and Operations in the Middle Grades,* edited by J. Hiebert and M. Behr, 93–118. Reston, VA: National Council of Teachers of Mathematics and Lawrence Erlbaum.

Louis, K. S., S. Kruse, and H. Marks. 1996. "Schoolwide Professional Community." In *Authentic Achievement: Restructuring Schools for Intellectual Quality,* edited by Fred Newmann and Associates. San Francisco: Jossey-Bass.

Marzano, R. J. 2000. *Transforming Classroom Grading.* Alexandria, VA: Association for Supervision and Curriculum Development.

McDonald, J. P., N. Mohr, A. Dichter, and E. C. McDonald. 2003. *The Power of Protocols.* New York: Teachers College Press.

National Council of Teachers of Mathematics. 2000a. *Lessons Learned from Research.* Reston, VA: National Council of Teachers of Mathematics.

———. 2000b. *Mathematics Assessment: A Practical Handbook.* Reston, VA: National Council of Teachers of Mathematics.

———. 2000c. *Principles and Standards for School Mathematics.* Reston, VA: National Council of Teachers of Mathematics.

National Research Council. 2000. *How People Learn: Brain, Mind, Experience, and School.* Washington, DC: National Academy Press.

———. 2001. *Adding It Up: Helping Children Learn Mathematics.* Washington, DC: National Academy Press.

Newmann, F., and G. Whelage. 1995. *Successful School Restructuring: A Report to the Public and Educators by the Center for Restructuring Schools.* Madison: University of Wisconsin.

Parke, C. S., S. Lane, E. A. Silver, and M. E. Magone. 2003. *Using Assessment to Improve Middle Grades Mathematics Teaching and Learning.* Reston, VA: National Council of Teachers of Mathematics.

Peck, J. M. 2005. *Improving Adolescent Mathematics: Findings from Research.* Portland, OR: Northwest Regional Educational Laboratory.

Pomerantz, H. 1997. "The Role of Calculators in Math Education." Report prepared for the Urban Systemic Initiative/Comprehensive Partnership for Mathematics and Science Achievement Superintendent's Forum, Dallas, TX.

Rasmussen, C., E. Yackel, and K. King. 2003. "Social and Sociomathematical Norms in the Mathematics Classroom." In *Teaching Mathematics Through Problem Solving,* edited by H. Schoen and R. Charles, 143–54. Reston, VA: National Council of Teachers of Mathematics.

Reys, B. 1991. *Developing Number Sense in the Middle Grades: Addenda Series, Grades 5–8.* Reston, VA: National Council of Teachers of Mathematics.

Rectanus, C. 1997. *Math by All Means: Area and Perimeter, Grades 5–6.* Sausalito, CA: Math Solutions Publications.

Stein, M. K., M. S. Smith, M. A. Henningson, and E. A. Silver. 1999. *Implementing Standards-Based Mathematics Instruction: A Casebook for Professional Development.* New York: Teachers College Press.

Talbert, J. E., and M. W. McLaughlin. 1993. "Understanding Teaching in Context." In *Teaching for Understanding: Challenges for Policy and Practice,* edited by D. K. Cohen, M. W. McLaughlin, and J. E. Talbert, 167–206. San Francisco, CA: Jossey-Bass.

Thompson, V., and K. Mayfield-Ingram. 1998. *Family Math: The Middle School Years.* Berkeley: The Regents of the University of California.

Yackel, E., and P. Cobb. 1996. "Sociomathematical Norms, Argumentation, and Autonomy in Mathematics." *Journal for Research in Mathematics Education* 27: 458–77.